Astronomer
by Chance

THIS BOOK IS PUBLISHED AS PART OF AN ALFRED P. SLOAN FOUNDATION PROGRAM

ASTRONOMER BY CHANCE

BERNARD LOVELL

Basic Books, Inc., Publishers   New York

Library of Congress Cataloging-in-Publication Data

Lovell, Bernard, Sir, 1913–
    Astronomer by chance/Bernard Lovell.
        p. cm.—(Alfred P. Sloan Foundation series)
    ISBN 0-465-00512-8
    1. Lovell, Bernard, Sir, 1913– . 2. Astronomy—
Great Britain—History—20th century. 3. Astronomers—
Great Britain—Biography.
I. Title. II. Series.
QB36.L848A3 1990
520'.92—dc20
[B]   89-43097
                                             CIP

Contents

CONTENTS

Preface to the Series

The Alfred P. Sloan Foundation has for many years had an interest in encouraging public understanding of science. Science in this century has become a complex endeavor. Scientific statements may reflect many centuries of experimentation and theory, and are likely to be expressed in the language of advanced mathematics or in highly technical terms. As scientific knowledge expands, the goal of general public understanding of science becomes increasingly difficult to reach.

Yet an understanding of the scientific enterprise, as distinct from data, concepts, and theories, is certainly within the grasp of us all. It is an enterprise conducted by men and women who are stimulated by hopes and purposes that are universal, rewarded by occasional successes, and distressed by setbacks. Science is an enterprise with its own rules and customs, but an understanding of that enterprise is accessible, for it is quintessentially human. And an understanding of the enterprise inevitably brings with it insights into the nature of its products.

The Sloan Foundation expresses great appreciation to the advisory committee. Present members include the chairman, Simon Michael Bessie, Co-Publisher, Cornelia and Michael Bessie Books; Howard Hiatt, Professor, School of Medicine, Harvard University; Eric R. Kandel, University Professor, Columbia University College of Physicians and Surgeons, and Senior Investigator, Howard

PREFACE

Hughes Medical Institute; Daniel Kevles, Professor of History, California Institute of Technology; Robert Merton, University Professor Emeritus, Columbia University; Paul Samuelson, Institute Professor of Economics, Massachusetts Institute of Technology; Robert Sinsheimer, Chancellor Emeritus, University of California, Santa Cruz; Steven Weinberg, Professor of Physics, University of Texas at Austin; and Stephen White, former Vice-President of the Alfred P. Sloan Foundation. Previous members of the committee were Daniel McFadden, Professor of Economics, and Philip Morrison, Professor of Physics, both of the Massachusetts Institute of Technology; George Miller, Professor Emeritus of Psychology, Princeton University; Mark Kac (deceased), formerly Professor of Mathematics, University of Southern California; and Frederick E. Terman (deceased), formerly Provost Emeritus, Stanford University. The Sloan Foundation has been represented by Arthur L. Singer, Jr., Stephen White, Eric Wanner, and Sandra Panem. The first publisher of the program, Harper & Row, was represented by Edward L. Burlingame and Sallie Coolidge. This volume is the eighth to be published by Basic Books, represented by Martin Kessler and Richard Liebmann-Smith.

—The Alfred P. Sloan Foundation

The original manuscript of this book was too long for inclusion in the Alfred P. Sloan Foundation Series, and I am deeply indebted to John R. Walsh, former member of the staff of *Science*, for the sympathetic manner in which he has helped me to make the necessary deletions and adjustments.

—Bernard Lovell

Astronomer
by Chance

1

Introduction

Toward the end of July 1939, when I was a month short of my twenty-sixth birthday, I was called to the telephone. Those few seconds of conversation were soon to lead to entirely unforeseen changes in my career and my outlook on life. The call was from Patrick Blackett, my professor of physics in the University of Manchester. I was in the quadrangle outside the physics department in Manchester, inspecting a small van that I intended to load with apparatus and drive across France to the Pyrenees, where, from a high altitude, I could continue my research on the cosmic radiation from space. Blackett was somewhere in London. During the last two years he had often been in London. I was soon to learn that these absences had not been for purely academic or research reasons. There had been no cause to think otherwise—he always returned full of enthusiasm with new ideas and instructions for our work. But this telephone call was a peremptory instruction to abandon my plans for purchasing the van and on no account to commence my journey across Europe.

For many months we had lived in the uneasy balance between war and peace. The Munich agreement of the previous September had done little to lessen the unease, but we had continued to teach and carry on with our researches and to plan our lives as though a war would not have a cataclysmic effect on *our* plans. That brief telephone call shattered this pretense. Blackett had often talked about war. He spoke from experience—he had been in the Battle of Jutland—and now he spoke from a knowledge of the crisis that we did not then comprehend. His instructions were clear, and six days later, on 4 August, I left Manchester with little idea that, apart from brief visits, it would be six years before I returned.

No doubt there are many people who, fifty years later, could remember the circumstances under which they heard Prime Minister Neville Chamberlain's broadcast speech announcing that we were at war with Germany. It is not so much where I was on that Sunday morning of 3 September 1939 but rather what I saw as Chamberlain was speaking that monopolizes my own memory. In fact, I was on the Yorkshire Wolds, near the seaside resort of Scarborough, in the operations room of one of the giant coastal defense radars. A young girl in the uniform of the Women's Auxiliary Air Force was seated before the large cathode-ray tube of the radar receiver. Her important task was to report to the headquarters of Fighter Command near London any radar echo from an aircraft that appeared on her screen. Naturally I expected that as soon as the prime minister declared that we were at war with Germany there would be a massive bombing raid. Indeed, many radar echoes appeared on the cathode-ray tube, but to my surprise the girl made no effort to report this through her telephone headset to Fighter Command. Why not? Ah, she responded, these are not radar echoes from aircraft; they are at the wrong range, and they are very short-lived. We were told they were radar reflections from the ionosphere.

I was not an ionospheric physicist, but I knew the fundamental facts that the ionosphere was the ionized region encircling the earth at a height of a hundred or more miles and that the reflection of radio waves from the electrons in that region made radio communication possible around the world. Cosmic-ray particles of high energy colliding with the earth's atmosphere also produced showers of electrons, and it was these cosmic-ray showers that I had been studying in Manchester and had proposed to investigate from the high-altitude Pic du Midi Observatory. So at the moment of the declaration of war the idea was born that radar might provide an entirely new and important technique for investigating cosmic-ray showers of high energy, but I had not the slightest inkling that this idea would eventually have a profound effect on my career.

Just over six years later a few trailers of army radar equipment were towed into a muddy field at a place called Jodrell Bank, about twenty-five miles south of the city of Manchester. With these I hoped to investigate this idea that radar could be used to study the high-energy cosmic rays from space. A major theme of the story that follows is not how this idea was realized but how the search led to completely new avenues of scientific exploration, to the construction of a massive new form of telescope for the investigation of the universe, and to a complex involvement in international affairs.

The Years of War

In the six years separating the sight of the radar echoes on the cathode-ray tube from the arrival of the trailers at Jodrell Bank, I was involved in the caldron of World War II. In the beginning I was an innocent, thankful to be allotted tasks concerned with radar in fighter aircraft and the defense of the country. The innocence and the idealism were soon shattered by the bombs that destroyed our cities, by the bullets that ripped into our research laboratories, by the air battles overhead, and by the death of colleagues. France had fallen; from our laboratory on the cliff edge we could look across the English Channel to where the Germans were preparing for their assault on England. At night, fearful of a parachute descent, we would drive our most secret equipment miles inland. After two years I was ordered to form a group to make a new radar aid so that our night bombers could find their targets in Germany. The task was accomplished against frightful odds in a time scale that, after forty years of peace, I still find almost inconceivable. In 1943 the bombing raids that destroyed Hamburg in a great fire storm and that tore the heart out of Leipzig and Berlin were guided by the Pathfinder Force using this equipment. In the heat of war, when life and death and the survival of the civilized world were at stake, that was success, but in my later years I have sought consolation in one of the recurrent ironies of scientific endeavor. That

3

lies in the narrowness of the dividing line between good and evil in that endeavor. This technical invention was designed to destroy Hitler's cities. By an accident of fate and timing, used in a different command and in a different role, it created the tactical situation that led to the destruction of the U-boats, the saving of merchant shipping, and the salvation of the people of the United Kingdom from near starvation.

In its many forms for defense and attack, radar was a vital factor—first in the survival of the United Kingdom against the German assault and then in the contribution to the Allied victory. But it did not make this vital contribution because it was a better radar than that of the enemy or because the Allied scientists were superior to those in the Fascist countries. The success arose from far more subtle reasons—the intimate collaboration between the operational groups and the scientists. Once I heard a frustrated air marshal complain to the head of our establishment, "The trouble with your lot, Rowe, is that your lowest-grade technician won't turn a screw unless we explain to him the tactical situation in the Middle East." It was not a great exaggeration. If the operational need was made clear and the scientific solution could be glimpsed, the question was never "How much will it cost?" but "How soon can we have it and what help do you want?" At a critical stage in the development of the V-2 rockets, Hitler imprisoned Wernher von Braun because of a rumor that he was interested in space flight and not in producing a military weapon. Such a lack of understanding and contact was inconceivable in the Allied scientific war.

The establishment to which I was directed shortly before the outbreak of war had been developing the coastal defense radar chain and pioneering forms of airborne radar. It was staffed with scientists and technicians who were members of the government civil service accustomed to regular hours of work and surrounded by regulations and conventions, so that even to purchase a screwdriver required a complex organizational procedure. Suddenly this group of a hundred or so was inundated with scientists from the universities like myself, accustomed to working when it was needed and unable to understand why it was necessary to order a simple tool instead of walking into a shop and buying one. I see now that the assimilation of these newcomers into a coherent whole to form one of the great wartime establishments was another major factor in the success of the Allied scientific wartime effort. Almost all were young men of exceptional brilliance who had developed the most advanced electronic tools for their own research disciplines.

In the first winter of the war a young man of my own age was sent to help me on an ice-bound airfield in South Wales. He was from the physiological laboratory in Cambridge, but I discovered within minutes that I had no need to explain to him how to use an oscilloscope. He had developed electronic equipment of great sensitivity to study the transmission of impulses along the nerves of a giant squid. He was Alan Hodgkin, who soon became primarily responsible for the first centimeter airborne interception radar. After the war he was awarded the Nobel Prize in physiology and became President of the Royal Society and Master of Trinity College, Cambridge. Similarly Andrew Huxley, also a physiologist, in 1942 showed me in five minutes how to obtain data about the fading of radar echoes when flying in an aircraft that had been puzzling me for days. He simply pulled a camera from his pocket, held it in front of the cathode-ray tube, grasped the end of the roll of film, and pulled it through so that the echoes that were imaging several hundred times a second were neatly separated on the film. Like Hodgkin, he also became a Nobel laureate after the war, President of the Royal Society, and Master of Trinity.

I mention Hodgkin and Huxley because I worked with them, but they were only typical of the galaxy of talent that erupted in the wartime establishments. They left behind them the pleasant environment of the university, with its freedom of research, and accomplished remarkable feats of scientific development often under hazardous and chaotic conditions.

The Return to the University—P. M. S. Blackett

Almost without exception we did return to our universities in 1945, but my own return was not a smooth continuation of the life and work of six years earlier. If it had been so, I doubt whether there would be a reason for this book. I returned to a

Manchester and a laboratory in which the reality was vastly different from the dream of peace that had sustained me during the heat of war. However, Blackett was returning, and like a magnet he pulled me against the influences that tried to retain me in the defense establishment or push me into industry. His name will occur frequently in these pages. He was the pivotal figure around whom my life and work revolved for eight critical years after the end of the war. When he decided to leave Manchester for London, in 1953, almost everything that followed was the result of our association in those immediate postwar years. Until his death twenty years later, he remained for me a father figure, although he was no longer responsible for my activities. When I was charged by the Royal Society with the task of writing his biographical memoir, the account seemed like a summary of the major scientific, political, and military events of the century—such was the scope of his activities and influence.

Blackett emerges as one of the most remarkable men of the twentieth century. He was renowned for his work as a physicist—first in nuclear physics, then in the study of cosmic rays, and finally in geophysics, where he produced the vital rock magnetic evidence about continental drift. He was an acute military strategist, a political figure of great significance in the postwar developments in the United Kingdom, and a passionate worker for the underdeveloped countries. That is the man whom I first met in 1936 as an unsuccessful applicant for a research post and who subsequently was to play a critical part in my postwar career and in the story that unfolds in this book.

Jodrell Bank

Today Jodrell Bank is one of the world's major observatories for the study of the universe. Its instruments dominate the Cheshire countryside for miles around, and every week thousands of visitors are attracted by the exhibits of its work. The contrast with my derelict trailers of ex-army radar equipment towed into a field there late in 1945 is remarkable. This transition from the relative freedom that I enjoyed immediately after the war to the control of major projects by committees involved a battle that raged around me for twenty years. That I was nearly a victim of this battle is attributable to the coincidence that I happened to be one of the first scientists in the postwar era to plan the construction of a massive scientific instrument—a radio telescope. Although I had no idea of this in the 1950s, I was involving myself, the universities, and the government in the initiatory problem of what became known as big science. Today the problems of big science are dominant factors in the national and international planning of scientific research. The modern instruments of space research, astronomy, and nuclear and atomic physics are vastly expensive and require large teams of engineers, technicians, and scientists. They are constructed, controlled, and operated as national and international facilities. Jodrell Bank remains almost unique in that it is a big-science facility owned and operated by a university. But the university alone could not have afforded the cost of the telescope. When the dust had settled in 1960, it had become a tripartite enterprise—university, government, and industry—with continued funding from university and state on a scale unimaginable in the prewar era. A new kind of science had evolved in which research depended on the collaboration of scientists, technicians, and engineers.

This story is, in the broadest sense, simply one example of how the nature of the physical sciences changed from the prewar to the postwar era. The scientists who emerged from the six years of war were radically different from those who worked in the seclusion of their laboratories earlier in the century. In 1939 we had been injected into a new kind of science, entirely different from that pursued alone at a laboratory bench. We had to learn to work in the disciplined world of the armed services and the industrialists, with massive installations—with airplanes, guns, and ships—and in teams of scientists and technicians. When the war ended, we may well have tried to shelve these associations and attitudes, but they had become ingrained in our outlook and were to be fundamental to the emergence of the projects of big science. Jodrell Bank was just one of these projects. It may not have been the first, but it was the first that was so obvious and the first that raised the question of cost-effectiveness in science to a national consideration.

Many years after the major conflicts over the instrument now known as the Lovell telescope had become quiescent, Sir Fred Hoyle wrote,

> Lovell suffered a good deal of criticism over this near doubling of the original estimate, criticism which he took well in public but evidently felt to be unjustified in private. With justice or not one can ask? And one could fill a book in attempting to answer the question. Suffice it to say here that Lovell may well prove to be the last of the great entrepreneurial scientists, in a style that does not mix easily with the modern bureaucratic organization of science through school, university and research generally. No real matter to bureaucracy that Lovell's expenditures were always highly cost-effective in terms of scientific productivity. Scientific productivity in our present research council-controlled system is no longer the real name of the game. First priority goes nowadays to maintaining the organizational structure of science, without much regard being paid to whether the structure we are presently saddled with is optimum in terms of productivity, or even within remote sighting distance of being optimum.*

Some may ask why I did not abandon the project of the telescope. Clearly there were many occasions when I might have felt justified in doing so and retiring to the calm of the unobtrusive laboratory workbench. I do not think such a retreat ever occurred to me as a possibility. There may well have been an element of obstinacy, but there was also something deeper than that—an instinct that it was vitally important for science to build this telescope. As the 1950s unfolded, another instinct developed that international issues outside the localized domain of research were at stake. The instincts were justified in a manner that I could not possibly have envisaged. I cannot explain the source of such instincts. They are an obscure part of the individual and his scientific endeavor.

The New Astronomy

The trailers that were towed to Jodrell Bank late in 1945 contained radar equipment. That is, there was a transmitter that, fed

*Sir Fred Hoyle, *The Observatory* 108 (1988):58.

into an aerial system, transmitted pulses of radio waves at a rate of several hundred per second. This particular system had been used to detect enemy aircraft for the army anti-aircraft guns during the war. When an aircraft passed through the beam of the radar, some of the energy in the pulses was scattered back to ground level and picked up by another aerial, connected to a receiver in a second trailer.

My idea was to use this system to detect the energy scattered back to the earth by the electrons formed by energetic cosmic-ray particles from space that gave rise to a cluster of electrons when they made collisions in the high atmosphere of the earth. I never succeeded in this experiment, partly because my attention and interest were diverted to the radar echoes that turned out to be from the ionized trails of meteors (shooting stars) that produced a long column of electrons as they evaporated in the atmosphere a hundred miles above the earth. We had a valuable new technique for investigating these meteoric phenomena, and for several years the majority of the research papers published from Jodrell Bank concerned meteors. Later, with more powerful transmitters and more sensitive receivers, this radar technique was used at Jodrell Bank and elsewhere to study the moon and the planets. It became known as *radar* astronomy.

Our attempts to build a larger receiving aerial to get even more sensitivity soon led us to a new kind of astronomy in which only the receiving part of the system was needed. That is, we became concerned with the detection and investigation of the radio waves that were reaching the earth from sources far away in the universe. After a few years this new science became known as *radio* astronomy, and that subject now completely monopolizes the research work at Jodrell Bank and many other observatories in the world.

When an aerial system or radio telescope is used in the receiving mode, the signals it receives from space are very weak because of the great distances of the source. But the sources in the universe that generate these radio waves are enormously powerful. The radio waves from objects in the universe that are billions of light-years distant can be received on earth without much difficulty. In fact, the aerial and receiver of an ordinary television set could be used to make a crude map of the strength of the radio waves emanating from regions of space that are millions of light-years distant. By contrast the range of a radar system using the same radio telescope as a transmitter and receiver is limited to a matter of light-minutes. The sun is eight light-minutes distant from earth and the nearest star about four light-years, and the modern radio astronomical systems penetrate billions of light-years into space.

The postwar discoveries revealed that unsuspected forms of energy production existed in the universe. Electrons moving at high velocities in magnetic fields generate energy in the radio spectrum immensely more powerful than that of the thermal radiation at radio wavelengths. Although this synchrotron process was recognized in the 1950s to be of great significance in the universe, few astronomers today would feel confident that the major energy-producing processes are understood. The contemporary discussions of black holes and gravitational collapse lie at the heart of the subject, as does our understanding of the large-scale structure and evolution of the universe. In the 1930s Edwin Hubble thought that the 100-inch optical telescope could penetrate 140 million light-years into space. The discoveries using radio telescopes have stimulated the extension of this penetration so that the collaborative researches of the optical and radio astronomers deal with penetrations into space of many billions of light-years and take us back in time to the epoch when the universe was in the earliest stages of expansion, some ten to twenty billion years ago.

No one yet fully understands why the galaxies or other objects in the universe are often such powerful generators of radio waves. Indeed, the question lies deep in the modern science of astrophysics, and the unfolding of this story in the years since the end of the war is the story of the development of a new astronomy in which Jodrell Bank has been immersed.

The Evolution of Jodrell Bank

In 1945 the university authorities gave me permission to station the trailers of radar equipment in the grounds of the botanical department at Jodrell Bank for two weeks. This book focuses on the extension of these two weeks to more than forty years and on the unplanned development of this new science of radio astronomy. A major restrospective lesson of these four decades is that research in science is very much an act of faith.

Large sums of money are today spent on studies, reviews, and committees in order to make sure that a scientific proposal is worthwhile and that it will succeed. Jodrell Bank would never have evolved or survived under such scrutiny. It is true that I had to face enough committees, boards of inquiry, and interference from anonymous people in official positions. At the moment of success of the telescope I had the unique experience for a scientist of suffering the threat of imprisonment. This sequence of events occurred because I wanted to build a new type of telescope in a Britain still ravaged by the effects of the war, and at a cost that would absorb a major slice of the budget available for fundamental research. However, it remains a surprising feature of the whole episode that the telescope that was the source of the troubles became one of the world's major instruments for the exploration of the universe.

In the beginning my targets were those of a physicist, not of an astronomer. I became an astronomer by accident because the instruments I built as a physicist became important by chance in this evolving science of radio astronomy. In the papers and memoranda of the late 1940s arguing the case for the massive 250-foot aperture telescope, the possibilities for its use in radio astronomy appear almost as a sideline. Decades later the telescope is not only wholly involved in radio astronomy but also studying objects that had not even been discovered, or whose existence in the universe was not suspected, at that time—radio galaxies, quasars, pulsars, hydrogen clouds, masers, and the relic radiation from the primeval state of the universe.

The trailers of redundant radar equipment in 1945 symbolized the end of a world war and of a whole range of military techniques. Twelve years later, in October 1957, the first Soviet Sputnik marked the beginning of a new phase of military technology. The rocket that launched Sputnik was the carrier rocket of the world's first intercontinental ballistic missile. By a strange irony

Sputnik saved me and Jodrell Bank from extinction. The telescope, in the final stages of preparation for astronomical use but heavily burdened with debt, was transformed almost overnight into the world's most powerful radar system. In the subsequent years of the cold war we were often sandwiched between the United States and the Soviet Union because in the new frontiers of space the telescope became a critical instrument in the developing military space technology of the major powers. For another decade I was once more astride the dividing line between the good and the evil in scientific endeavor. After Neil Armstrong stepped onto the moon, in 1969, these tensions faded, and we were free at last to concentrate on the problems of astronomy.

The Personal Factor

During my life the popular image of the scientist has undergone a transformation. In the early years of the century he was often pictured as an absentminded recluse, far removed from the problems of the world. Indeed, this was often true, and scientific discoveries rarely attracted public attention. The work of a few of the great scientists early in this century began to penetrate the public consciousness. An example of this was Einstein and the theory of relativity, particularly when the observations in 1919 of stellar positions during a total solar eclipse were hailed as a verification of his general theory. That certainly reached the public; in fact, one of the first stirrings of my interest in science was through the mention of Einstein's theory of relativity in a sermon preached by a visiting parson in our isolated village when I was a young boy. However, the fateful implication of the earlier special theory of the equivalence of mass and energy attracted little general attention. And the work of Ernest Rutherford on radioactivity and the disintegration of the atom was presented to the public as being of interest only to specialists. As a student I heard Rutherford lecture on the recent splitting of the atom by J. D. Cockcroft and E. T. S. Walton and issue his authoritative denial that the energy locked up in the atom could ever be released for practical purposes—yet the dropping of the atomic bomb on Hiroshima and Nagasaki lay only twelve years ahead.

ASTRONOMER BY CHANCE

The Second World War was the event that transformed science and scientists. The governments of the major powers had seen the relevance of science to warfare. Whereas science had previously been a major intellectual pursuit, it now became a relevant discipline in the search for economic and military strength. The heads of state appointed scientific advisers, and there began a massive diversion of scientific talent to the support of industrial and technological developments and to the administrative affairs of science. In the military sphere the possession of nuclear arsenals has led to the perilous balance of power—not to guarantees of superiority. As a scientist I have lived through and been a part of this transformation, which has within a generation eroded the differences that formerly distinguished the scientist from his neighbor.

The advance of science has given neither absolute material power nor absolute knowledge of the universe and of man's relation to the cosmos. The arrogant belief prevalent in my youth in the ultimate power of science to explain the universe and all that lies within has passed with the experience that the search for solutions uncovers ever deeper mysteries. Today the world of the scientist is less hostile to the view that science does not have the sole prerogative of acquiring ultimate knowledge.

Perhaps as a child I was fortunate in being born into a household and an isolated village community that fervently believed in God and the devil and heaven and hell. I see now that this good fortune continued in my youth when the iconoclastic influences of university life were counterbalanced for me by the discovery of Alfred North Whitehead's Lowell Lectures at Harvard, published as *Science and the Modern World.* This relevation of the interactive and transdisciplinary nature of science, philosophy, and theology stimulated an interest in our conception and relation to the universe that survived the impact of the war. An intensification of interest in these fundamental issues came in 1958. At a moment when I was facing the real threat of imprisonment because of the complex financial and political issues raised by the successful completion of the telescope, I was stirred from the lethargy of hopelessness by an invitation to deliver the Reith Lectures for the British Broadcasting Corporation.

13

For the first time after the experiences of the war, I was forced to give expression to the disturbing conflicts of my life—the reconciliation of the religious interests of my childhood with the implications of contemporary science, and the shattering of youthful idealism in the mass destruction of war. These six lectures, published as *The Individual and the Universe*, marked a turning point in my own effort to resolve the conflicts that face every scientist.

There was no science in my familial origins or my early childhood. I had no instincts or ambitions to become a scientist until they surfaced with a degree of inevitability while I was attending, during my schooldays, a public lecture on the electric spark. As with so many other events in my life, I cannot explain why this happened, but when it did there was no turning back. By choice I became a scientist, and by chance I became an astronomer.

2

The Years of
Boyhood and Innocence

The cities of Bath and Bristol, in the west of England, are separated by less than twenty miles. The former is an ancient Roman city of great beauty, beloved by tourists and unspoiled by the industries of the twentieth century. Bristol is quite different, owing its initial wealth and prosperity to the merchants who traded over the seas hundreds of years ago and, today, thriving on the industries of the modern age. In my youth the river Avon penetrated to the very heart of the city, so that ships were moored beside the pavements. After the destruction of the city center by the German bombs of World War II, the city fathers decided to extend the central area, and today the casual visitor is unaware that the city was once a thriving port from which Cabot sailed in the fifteenth century to discover the New World across the ocean.

As I write now in my seventh decade, I recall with some surprise that nearly a third of my life has been spent in this pleasant environment of Bath and Bristol—for an ancient milestone only a few hundred yards from my birthplace recorded that the distance to Bath was seven miles and to Briftol (sic) seven miles. The traffic now roars past the place where that milestone once stood, and it is hard to imagine that I once bowled a hoop along that road and played fivestones on its surface without fear of injury, because there was nothing more menacing than the occasional bicycle or pony and trap.

Although so close to Bath and Bristol, the village of my birth and upbringing was quite isolated. Getting to the station for the nearest steam train that took one to Bath in one direction and Bristol in the other involved a two-mile walk. The terminus of the more frequent electric trams to Bristol was two miles in the opposite direction. The postwar expansion of the city of Bristol has engulfed that village of Oldland Common, and the green fields in which I picked the cowslips and flew my kite are now covered with houses or shopping centers. The railway station, the trains, and the electric trams have vanished in this day of the ubiquitous automobile.

Even so, such villages seem able to retain some of their ancient identity. The house in which I was born on 31 August 1913 still stands, a general shop, curiously bearing the name of Adams as it did on the day of my birth. The Adams of today does not bear any close relation to my widowed grandmother Adams, who was the owner of that business in my boyhood days. She wrote details of purchases with a scratchy pen in a large ledger because in the days of the First World War and the 1920s few of the villagers had ready money to pay for their goods. Rather than see them starve or suffer hardship, my grandmother allowed them credit, which was rarely redeemed. She lived until I was eighteen, and I remember the astonishment of my parents when the extent of this unredeemed credit was discovered—a situation that no modern businessman could tolerate or contemplate. One of her daughters, Emily Laura Adams, was my mother. She lived to be over ninety, when she could still read a newspaper without spectacles. In her younger days she was an intensely practical and handsome woman—a photograph reveals her as the captain of the Kingswood Ladies Cricket Team in the days before the First World War, when such activities for her sex must have been rare.

The single-story, specially built nearby house to which we moved when I was a boy still stands, and the adjacent field on which cows grazed is now the site of a modern school. I am glad that the school bears my name, because it is the replacement of the old village school where I learned to multiply and divide without the assistance of the pocket calculators of today.

Astronomer by Chance

The village communities of the early years of this century formed worlds of their own, and marriages were rarely made with "foreigners" from other communities. In fact, Emily Laura Adams married Gilbert Lovell, a local tradesman whose business was a mere hundred yards from her mother's, and I was their only child. The families of the Adamses and the Lovells were by far the most extensive in that village and influenced almost every activity, be it musical, religious, social, or sporting. Apart from the normal village schooling, my father was self-educated, and in my later years I realized that this was a remarkable achievement. He was undoubtedly the most widely read person in the village. He had more knowledge of English grammar and language than I have ever achieved. When I was forty-four years old, as will be related in this book, I delivered the Reith Lectures of the British Broadcasting Corporation. After each lecture my father would write or telephone, concerned not about the substance of the lectures but about my pronunciation, the structure of my sentences, and especially my delivery. I think that under other circumstances he would have been a fine orator. He was one of the most truly religious persons I have ever encountered. The Methodist chapels were organized in circuits of some dozen chapels, and each circuit was in the charge of an ordained minister assisted by a number of lay preachers. As one of these lay preachers, my father would preach to crowded Methodist chapels and Sunday schools four times every Sunday, often traveling miles on his bicycle through snow and ice in winter to do so. His sermons were based on a profound knowledge of the Bible, and he used the imagery and language of the authorized King James text with powerful effect. The reprimands of my upbringing were biblical quotations delivered unerringly from memory, the chapter and verse being quoted. On summer mornings soon after sunrise, I would often be awakened by my father's voice and would look sleepily out of the window. I still retain the picture of him beyond the lilies practicing again and again the delivery of a single sentence from the Bible with various degrees of emphasis until he was satisfied with the interpretation. In later years it was my good fortune to become friendly with various professional musicians.

17

When they stayed in our house, I would again hear early in the morning the constant repetition of a few bars that so reminded me of my father's rehearsal of the spoken word.

The ancestry of my parents can be traced back in the registers of the local parish churches for a few hundred years, but there is nothing in these records to suggest that a scientist would one day emerge from those various unions. Adventurers, yes. Lovells sailing from Bristol were among the first white settlers in New Zealand and elsewhere. Ruffians, too. The eighteenth-century journals of John Wesley pay tribute to the colliers of Kingswood, Bristol, among whom our ancestors were numbered, as a people famous "for neither fearing God nor regarding man: so ignorant of the things of God, that they seemed but one remove from the beasts that perish; and therefore utterly without desire of instruction, as well as without the means of it."* It is fortunate that the drunkenness that killed many of these forebears was largely excised in the age of my grandfather, who learned to read at the age of eighty and founded the local Sunday school.

The emergence of technical and scientific interests in a child with such ancestry is indeed curious. My parents certainly encouraged my embryonic interests. Music and cricket were understandable. Both activities were inherent in the families of the Lovells and the Adamses. One branch of my father's family founded its own military band of Lovells, and there were enough good cricketing Adamses in the village to challenge the village cricket eleven. When I was a young boy, I was once summoned urgently to the cricket field. The Adams team was one short— "But you will do; your mother was an Adams." Two of my uncles were the stars of the annual village operetta, and my attachment to the organ relates closely to two others. One of these uncles, Alfred Adams, gave me my first organ lessons, and I deputized for him at an early age as organist at the Sunday services. The other uncle would drag me as a child through the Bristol streets to hear visiting virtuosi perform on the organ of Colston Hall. But they were not scientists, nor were they technically inclined.

*From a letter to a "Mr. D.," 27 November 1739, in *John Wesley's Journal*, abridged by P. L. Parker (London: Isbister, 1903), p. 68.

ASTRONOMER BY CHANCE

Since I was twenty-six, when the war began, my career has been involved in various ways with the techniques of radar and radio, and it may be more than a coincidence that my earliest technical interest as a child was aroused by the talk of the new invention of the "wireless." At some time in the early 1920s two tall wooden poles were erected in the garden about twenty yards apart. Between them a single wire was supported by insulators at each end and a down lead came through the window frame via an ebonite insulator. This was connected to a small box on which glowed two thermionic valves; by donning headphones and manipulating various dials on the box, one could hear faint noises or music. Because no electricity was supplied to the village, the device was operated from batteries. Mystified and enthralled by this apparent miracle, I soon learned how to make my own simple receiver using only a small crystal on which a piece of wire would be scratched until faint noises would be heard in the headphones. The progression to the construction of more ambitious instruments was natural, and by my eleventh year, when I was transferred to a secondary school, I was a regular reader of the weekly wireless journal and had built equipment that would transmit as well as receive. In my early teens that and the game of cricket formed my world. The secondary school had been opened only a year or two earlier, and I was one of the first pupils. It was two miles distant, and I would cycle there every morning with my cousin Harry Lovell, who had also been one of the first pupils. He became an architect, and in later life our paths crossed occasionally over vital issues concerning housing developments in the vicinity of the Jodrell Bank telescope. It seems most unlikely that either of us would have had a professional life as architect or scientist but for the fortuitous opening of that school when we were children.

These pursuits monopolized my interest to such an extent that I was a poor pupil academically. The candles and oil lamps in our house were replaced by a small petrol-driven electric generator plant, and I became inseparable from the oily machinery, the dynamo driven by a canvas belt from the engine, the row of accumulators, and the switches, ammeters, and voltmeters on the switchboard. At the age of fourteen I was near the bottom of the

19

class but probably had as much practical knowledge of radio and electricity as anyone for many miles around and used the tools of the trade to construct ever more ambitious equipment. I had one desire—to leave school and expand my father's business into these new and fascinating activities.

In the early months of 1928 that was by far my most likely destiny. It did not occur because of a sequence of chance happenings that any operational analyst would now consider to be highly improbable. The first concerns the headmaster of the school. A retired army major, M. J. Eaton was a strict disciplinarian. He did not forgo the use of the cane; across the knuckles for minor offenses, but more severe punishment before the assembled school for offenses he thought might bring the school into public disrepute. He may have been wise, for he was in charge of a new school and anxious to establish a tradition of success. No doubt this anxiety for academic success led him to scrutinize the small number of pupils in the school with unusual insight. Although I ranked near the bottom of the class, he wrote on my report that I showed "distinct ability in mathematics and science," and his pleadings with my father to allow me to continue my studies and to sit for the school certificate examinations succeeded against my own desires for a more immediately practical life.

With some reluctance, in the autumn of 1928, shortly after my fifteenth birthday, I began to prepare for the examinations in the summer of 1929, believing this to be my final year as a schoolboy. But in that autumn three incidents within a few weeks changed my life. The school was not a boarding school, and every day I would cycle to my home, where I would spend tedious hours attempting to do the homework set for that day. The teachers maintained that twenty minutes on each of the three subjects set for that evening should be adequate time. Perhaps bright and attentive pupils could accomplish the work in that time scale. I was neither and would spend the entire evening on the homework when my spirit yearned for the wireless gadgetry that surrounded me. On one such autumn night a friend of my father's called to consult him on a religious topic. His name was Champion. I doubt if I ever heard his Christian name mentioned; such intimacies were reserved for the family circle in those days. In any event, Mr. Champion was a mathematics master in another school, and on that evening he showed me the elegant way to deal with a problem in dynamics that had utterly confused me for the past hour. To the astonishment of the master and the rest of the class the next morning, it appeared that I was the only boy to have solved the problem. Those few moments in the company of Champion had released an elementary mathematical insight and such confidence that for the rest of my schooldays mathematics was never a problem.

Soon after that evening my future as a scientist and not as a shopkeeper or businessman was determined. I had reached the stage, annoying to my father, when my interest in the Sunday services was maintained only when I was on the organ stool. In particular the sermons were often long and tediously evangelical in content, and I tended to read more interesting literature that I had smuggled to the security of the organ stool among my music. On one such Sunday evening the preacher was the ordained minister of that Methodist circuit of churches. I wish I could recall his text, but my attention was already elsewhere until suddenly I heard him talking about Einstein, relativity, and the four dimensions of space-time. I doubt if the congregation understood his sermon any more than I did myself, but he had opened another kind of world, a world that within days was to be revealed to me in dramatic fashion—the decisive event that determined my future.

The senior physics master in the school had arranged to take a group of the boys to a series of evening lectures given in the physics laboratory of the University of Bristol by Professor A. M. Tyndall. Many years later I discovered that these lectures on the "electric spark" had been prepared and delivered by Tyndall as one of the famous annual Christmas lectures for young people at the Royal Institution in London. In 1928 I knew nothing of the Royal Institution and had never been near a university. My introduction to the University of Bristol that evening was overwhelming. The physics department was in a new building, the large lecture room was magnificently equipped, and Tyndall produced electric arcs that ripped across the lecture room, discharges in glass vessels that changed color as they were evacuated, and demon-strations of the properties of infrared and ultraviolet light that I was never to forget. That lecture room held all the gadgets and devices that represented my idea of paradise, and suddenly I wanted desperately to become a student in Tyndall's laboratory. Three years and two examinations later, I was there.

3

The Years in the
University of Bristol,
1931–1936

In the autumn of 1931, a few weeks after my eighteenth birthday, I became a student in the University of Bristol. The physics laboratory that had so excited me during the chance visits of a few years earlier now almost completely encircled my life. Every day that magnificent lecture theater across which I had watched the great electric arcs in Tyndall's lectures became the place where I absorbed the foundations of physics and learned of the exciting developments of that era.

The carvings in the stone over the massive oak entrance doors of the laboratory acquired a new meaning. On the left, the single beam of light separated by the prism into the rays of the spectrum; on the right, three lines representing the deflection of an alpha particle in collision with the nucleus of an atom. Carved in stone, the lines were similar. In the world of physics on the other side of that entrance door they became symbolic of the change in our understanding of the natural world from the seventeenth-century prism of Newton to the twentieth-century atomic world of Rutherford.

Many years later when circumstances wrenched me away from that laboratory, I realized how fortunate I had been in my student days. The H. H. Wills Physics Laboratory had been built and endowed only a few years earlier through the munificence of the Wills family—the wealthy tobacco magnates of Bristol. Constructed like a fortress in the grounds of the Royal Fort, it had a spaciousness and air of permanence that were soon to disappear from the purpose-built laboratories of later years. The great tower seen from afar dominated the Bristol skyline, and to that landmark I made the daily journey from the unsophisticated village of my childhood. The village was still isolated, and the early-morning two-mile walk to a train that would deposit me in the center of Bristol no doubt maintained my physical condition through the next three years of days within the walls of the laboratory and ceaseless hours at night with the books from which I learned the foundations of classical physics and mathematics.

That autumn of 1931 marked a change in my life and attitude that was self-imposed. My passion for the game of cricket would have carried me into the university team without difficulty. In the early days of my university life the captain of the university eleven introduced himself to me. I was in the library attempting to elucidate a problem in integral calculus set earlier that morning by James Vint, one of the lecturers in mathematics. Silence was enjoined in the library, and the captain motioned me to the corridor. "You will, of course, join the university team, Lovell; we begin practice early in the New Year"—"Thank you, but no, I've come to the university to work, and to be a member of your team would be too much distraction"—"But that is a stupid reason; after all, Vint, who set you that problem, bowls for us." I was adamant, and the captain, K. E. W. Ridley, of splendid physique, departed, as mystified as I became myself over that self-discipline, when, three years later, with the examinations behind me, I began to play for the university, as well as for a Bristol midweek team and the village.

However, those days of release lay years ahead when in the autumn of 1931 another event of great significance occurred. For luncheon I would eat a few sandwiches and some fruit. If the day was fine, I would find a seat in the lovely grounds surrounding the Royal Fort. There on one mild autumn day I was accosted by a fellow student. Only six of us were in the honors physics class, and that one of these, Deryck Chesterman, should walk toward me on that day is one of life's intriguing coincidences. I offered him an apple, and soon we were talking, not of physics, but of music. Up to that time our lives had been quite different. I was a country boy delighted, but still overawed, to be in the university. He was a public school boy brought up in the city of Bath, of which his maternal grandfather had been a distinguished mayor. His natural assumption that he would proceed to the University of Cambridge had been dashed by the financial depression of those years, and as I looked up with admiration to the University of Bristol, so he regarded his fate as too provincial. Drawn together as if from opposite poles of the magnet, we soon became inseparable friends. Through student days, through war and peace, we remained so for the next half century until his lamented death in 1978. He was the only close male friend of my life, a friendship of the purest character cemented together first by an intense mutual musical interest and after a few years by the deepest of all influences on my life.

That influence had its origin in music. In those prewar days distinguished soloists would perform on a weekday afternoon in Bath in the historic surroundings of the Pump Room. Chesterman, as he was then known to me, persuaded me to forsake work one afternoon to attend one of these recitals. With the hospitality that came naturally to his family, he took me to his home for luncheon before the concert. I do not think I was expected, because a lanky schoolgirl at the table was ordered by her mother to get another glass. Her name was Joyce. She was fifteen years old and wore a Girl Guide uniform covered with every conceivable Girl Guide badge. Apart from the badges the only other thing I noticed was the speed with which she ate her luncheon and departed from the table—to return to a private school a few streets away run by her two maiden aunts, so I was informed.

For the next three years Joyce Chesterman had no significant part in my life. I have little recollection other than of work of these years of my undergraduate course. I became seized with a passion for physics. Some of my early textbooks are on the shelves now. They are bulging with neat insertions in my handwriting either elucidating or correcting the text. For some of the lecturers I must have been a great irritant. In fact, where electronics was concerned, my years of practical experience as a boy and my absorption in the few journals then existing undoubtedly placed me in a position better than that of the lecturers. My constant interruptions and corrections led the lecturer on one occasion to suggest that I give the lecture. However, that overkill on one subject nearly ended in my downfall in the final honors examinations. In one paper a question on electronics so absorbed my interest that I had no time to answer the required number of questions. As every examinee must know, a perfect answer to one question when four are requested can only lead to failure. That, coupled with an abysmal performance in the *viva* examination, where I could not give satisfactory answers to questions on thick lenses, nearly cost me a first-class degree. Twenty years later I encountered Professor William Curtis, who had been the external examiner. I thought he would have completely forgotten that he once examined me on a summer's afternoon in 1934, but he remembered the occasion as well as I did. Curtis confirmed what I had long suspected, that it was the pleadings of Tyndall that saved the day for me.

Tyndall's public lectures had turned my mind to science, and I had become one of his favorite students. In fact, I had performed poorly in the *viva* examination simply because I had expected to be examined on some of the minor pieces of research I had already carried out as a student and not on the facts of elementary physics that could be found in a textbook. I had become so absorbed in physics and the processes of discovery that for me the vacations appeared as a barren land, and I had begged to be allowed to help in one of the many researches then proceeding in the laboratory. Such a request was greeted with astonishment, but I was allocated a few simple pieces of research and so learned the thrill of discovering new facts about nature years before I might otherwise have done.

Recently I was asked to comment on the differences between the undergraduate teaching of today compared with that of the interwar period. I was then teaching a class of 120 honors physics students and became aware of the gulf that separated these students from those of my own student days, when I was in a class of only 6. The intimate contact between teacher and student had vanished. The scope of physics had extended beyond the imagination of the 1930s. Our courses encompassed almost the whole of physics, but the language of solid-state, nuclear physics and the world of computers would have had no meaning in the 1930s, and the selection of courses forced upon the student of today inevitably poses a specialization unthinkable in my youth.

During my three years as an undergraduate the revolution in physics had commenced. Then the number of known fundamental particles of nature—the electron and the proton—was increased by the discovery of the positive electron and the neutron. In my final year the lecturer would appear before us in a state of excitement carrying a copy of *Nature* with such revolutionary news items as the discovery of artificial radioactivity or the disintegration of the atom by protons accelerated to high energies. For two years J. E. Lennard-Jones, the professor of theoretical physics, had covered the board with the Maxwellian equations of classical physics. Then came the youthful Nevill Mott, in later years the Cavendish Professor of Physics at Cambridge and a Nobel Prize winner. He wrote different kinds of equations—the equations of the new physics of quantum mechanics, Erwin Schrödringer's wave equations, and those of Werner Heisenberg's uncertainty principle. The superb clarity and beauty of these lectures remain a vivid memory. The determinate and certain world of common experience disappeared before our eyes as the probability functions of the wave equations became the new world of science that we were to inherit.

The books of Arthur Eddington and James Jeans describing the universe then being revealed through the 100-inch Mt. Wilson telescope have, for me, never been equaled by any detective story. That is how I first learned of Hubble's discovery of the extragalactic nebulae, of the vast extension of our penetration into space, and of his crucial argument for an expanding universe. I read without any thought that one day a strange chain of circumstances would lead me to study the universe with techniques and methods that were then unknown. I also tried and failed to

27

understand the *Principia Mathematica* of Bertrand Russell and Whitehead but was thereby led to read Whitehead. His *Science and the Modern World* revealed an understanding and interaction of science, philosophy, and theology that stimulated in me an enduring interest in our conception of and relation to the universe. The beautiful prose of his *Adventures of Ideas* enchanted me then, as it does today.

In this manner the simple and unsophisticated world of my village upbringing was transformed in those three years of my undergraduate days. So, too, was Europe being transformed as Hitler and Mussolini began their rise to power, but of that, and of the possible consequences to the world, I was then blissfully unaware.

The Postgraduate Years, 1934–1936

That innocence did not survive for long. Tyndall had no doubt that I would want to become a research student in the laboratory to work for the Ph.D. degree when my first-degree course had been completed. He was correct—and I had never visualized any life other than the pursuit of original research in physics. In those days the Department of Scientific and Industrial Research (DSIR) was the avenue through which the government supported fundamental research in the university. With Tyndall's recommendation I obtained a maintenance grant, which the DSIR offered to enable suitable students to proceed to postgraduate research, and in the autumn of 1934 my life entered another phase. The maintenance grant was only £120 per year. Even allowing for the subsequent inflation, this was a minor grant compared with the demands of students in later years, but in those days life was relatively simple and unsophisticated, and my parents, with whom I continued to live, made sure that the rigors of life did not endanger my health.

The life of a student is governed by a rigid timetable of lectures and examinations. That of the experimental research worker is controlled by more subtle factors and especially by the nature of the equipment with which he works. Often there may be long fallow preparatory periods and then periods of intense activity, when life's routine is determined by the apparatus and not by the clock. One must have freedom, but it is a freedom bounded by an iron discipline. It was my good fortune to have five years of this type of freedom before I was plunged into the wartime research establishment, where my comings and goings were recorded by the gatekeeper. The clock and the closure of the laboratory door militate against the essential attitudes to fundamental research, and in the Bristol laboratory of the 1930s the doors were never closed, and I soon learned that it might be more productive to be in the laboratory at midnight rather than at midday.

Nowadays, research in physics tends to demand large and expensive equipment, and research laboratories have become highly specialized. In the Bristol laboratory of the 1930s there was no such specialization on a single topic, but there was a wide variety of fascinating researches in progress carried out by individuals, perhaps assisted by one or two research students. Tyndall must have had faith in my ability because I was not placed in one of these existing groups. There was an interesting problem, it appeared, concerning the electrical conductivity of metals. An American scientist, Herbert E. Ives of the Bell Telephone Laboratories, had discovered that thin layers of alkali metals had a very much higher resistivity than the metal in bulk, and the literature about this phenomenon from a few investigators elsewhere was full of discordant results—both as to the thickness of the film where this anomaly became manifest and the reason this was so. That was to be my problem—to find how thick the film was when the anomaly appeared and why. I was to work alone but would be supervised by Dr. E. T. S. Appleyard, a senior lecturer in the laboratory. Appleyard had the misfortune to be afflicted with a spinal deformity and, in common with many whose activities are similarly restricted, had an encyclopedic knowledge of many subjects. It was an excellent arrangement. My years of experience throughout boyhood and youth with making my own radio equipment had given me the ability to construct and handle delicate equipment. Appleyard suggested the procedures, and I would translate them into the equipment on the bench. Soon I was surrounded by vacuum pumps, delicate gal-

vanometers, and a mass of equipment. For me it was a new paradise. My rapid progress was assisted by the good fortune that the
laboratory had one of the finest glassblowers in Europe—J. H.
Burrow—and like many other researchers in the laboratory I was
soon needing his help.

Johnny Burrow assessed his own priorities, and those he did
not like were low on his list. When Burrow discovered that I was
not afraid to do some minor glassblowing and that I could handle
lathes and the paraphernalia of the workshop with reasonable
skill, we became great friends and I went right to the top of his
unofficial priority list. Over the next two years Burrow produced
for me a series of very beautiful and delicate pieces of equipment
in Pyrex glass.

With this equipment I was able to deposit a thin film of one of
the alkali metals, rubidium, on a flat surface of Pyrex. Wire
probes inserted into this surface enabled me to measure the electrical resistance as the thickness of the deposited rubidium film
increased in a vessel from which the air was continuously evacuated. My first experiments gave the same discordant results as
those in the literature—that is, as soon as the deposited film
showed any electrical conductivity, the resistivity was anoma-
lously high compared with that of the bulk metal. Impurities in
the layer seemed to be a possible cause of the anomaly; in order
to allow me to test this, Burrow built me another piece of equipment and I built a large electrically heated furnace that could surround the whole glass system. With this I raised the temperature
of the system to just below the softening point of the Pyrex glass
and kept it there for several days and nights, with the pumps
maintaining a very high vacuum and with Burrow on the verge
of tears in case the product of his skill should be melted. After
this ruthless cleansing procedure was completed and the system
had cooled, the conductivity measurements were repeated. The
answer was clear: as soon as a thickness of only a few atomic
layers had been deposited, the resistivity became indistinguish-
able from that of the bulk metal.

Appleyard was cautious—it might be a coincidence. So more
demands on Burrow for a new glass system with which I could
deposit not only the atoms of rubidium but also a controlled

impurity. That was conclusive. I would deposit rubidium atoms on the baked and clean surface, and all would be well. When this procedure was repeated under the same conditions but with atoms of an impurity being deposited simultaneously, the anomalies would begin. Over two years I had repeated these measurements using cesium and potassium, and all gave the same results. Only under the most stringent conditions of high vacuum and purity could one obtain the electrical properties equivalent to those of the metal in bulk. During 1936 the results were published in *Nature* and in the *Proceedings of the Royal Society,* and in altogether too short a time I had completed the task set for my Ph.D. course. Nevill Mott had guided me to show how the results could be explained by a simple application of wave mechanics, and I should have been a happy young man.

My state of mind was not, however, a very happy one. Factors other than physics were disturbing my life. My innocence about the state of Europe had vanished. The laboratory was rapidly accumulating refugees from Germany. The room in which I worked was a large one, and it was shared by one of these young people from Germany whose inquisitiveness and interference I disliked. I was irritated by his talk of Hitler and his opinions so often underlined by newspaper placards. Even worse, there was Klaus Fuchs in a nearby room. After the war when he was arrested as an atomic spy and traitor, I felt no surprise. I remembered my earlier natural antagonism to him and my intense dislike of his probing questions.

By that time, too, I was deeply in love with Joyce Chesterman. In the autumn of 1934 when I began my postgraduate research, she had commenced a three-year teacher training course at Bedford. We had encountered each other at intervals since the day in 1931 when I first set eyes on the ardent Girl Guide, but now the lanky schoolgirl of 1931 had grown into a lithe and handsome young woman. Suddenly and inexplicably I knew early in 1936 that, of all my girlfriends, Joyce Chesterman was the girl I must marry. Before this could happen, there were to be many long and sad separations, and I would console myself by reading Whitehead: "Youth is too chequered to be termed happy. The memories of youth are better to live through than is youth itself. . . .

Youth is not peaceful in any ordinary sense of that term. In youth despair is overwhelming. There is then no tomorrow, no memory of disasters survived."*

In September 1937, a week after Joyce's twenty-first birthday and on the day she learned of the high honors gained in the college examinations, we were married. I write about this fifty years after the marriage service in Bath, and the reader of this book will search in vain for signs of rift in that early and mysterious bond that linked us so long ago. Neither the caldron of the war nor the tempestuous happenings in my later life have shaken the image of that ardent fifteen-year-old rightly adorned with the badges of prowess and achievement.

Throughout the two years of postgraduate life in Bristol, I satisfied two other longings suppressed during the undergraduate years. At last I played cricket for the university. The lovely ground at Coombe Dingle is one of many happy memories. Nearly fifty years later I went there again when the British Broadcasting Corporation was making a film of my life, and the magic was still there. The hard tennis courts on which I had played in the winter and the acres of grass on which I had bowled and batted with a youthful vigor remained unspoiled. At that time, too, I became a serious student of the organ. Deryck Chesterman introduced me to a close friend of his in Bath, Raymond Jones, known to us as Jonah. Later he was appointed the assistant organist at Bath Abbey and is now "organist emeritus." In 1935 he was the organist at St. Paul's Church in Bath. This church was destroyed during the war, but in the mid 1930s it had a fine organ and there I became Jonah's pupil. I had permission to practice on the organ in a church in Bristol near the physics laboratory. At lunchtime I would take some sandwiches and drive across the Clifton Suspension Bridge, spanning the Avon Gorge, to the church in Leigh Woods, and in that way I would practice the Bach that Jonah had assigned to me for the week.

That is how I lived in the years that Mussolini invaded Abyssinia and Hitler marched into the Rhineland. Neither these events nor the outbreak of the Spanish civil war in 1936 disturbed me

*Alfred North Whitehead, *Adventures of Ideas* (Cambridge, Eng.: Cambridge University Press, 1933), p. 370.

then as much as a talk with Tyndall in the summer of that year. My DSIR grant was for two years, and I had completed the research for my Ph.D. degree. I had no intention of leaving the pleasant ambience of Bristol, and Tyndall had obtained a Colston Research Fellowship for me. That was his marvelous way of screening from me a piece of advice that dismayed me beyond reason. On a lovely summer's day when I was thinking of nothing but the cricket match ahead, he told me that in his opinion it was time for me to leave Bristol to broaden my scientific outlook and to face greater competition than existed in his laboratory.

4

Blackett and Bragg—Manchester, 1936–1939

Tyndall's advice to me in July 1936 had been stimulated by an inquiry from P. M. S. Blackett, who wanted a young man to help with his cosmic-ray research at Birkbeck College in London, and at the same time W. L. Bragg, the professor of physics in Manchester, had asked him to recommend a young man for the post of assistant lecturer in Manchester. Reluctantly following Tyndall's advice, I applied for both of these posts and soon received telegrams from Blackett and Bragg asking me to come to London and Manchester to be interviewed on the same day at the same time in early August.

I had no desire or inclination to go to Manchester. I had spent my life in the West Country, and all that I had ever heard about Manchester appalled me. Even today I feel embarrassed at the arrogant manner in which I responded to that invitation from Bragg. It was to the effect that I could not attend at that time but would do so on the following morning, in the unlikely event that Blackett did not offer me the job in London. To work with Blackett would be a different matter. To the young physicist he was already a glamorous hero; having fought as a young naval lieu-tenant in the First World War, he had then worked in Ruther-ford's famous laboratory in Cambridge. His remarkable cloud chamber photographs of the collision of an alpha particle with the nucleus of a nitrogen atom and the tracks of the recoil oxygen isotope and a proton adorned every modern textbook of physics. In my student years his renown had surged still further when he discovered the tracks of a positive electron in the cloud chamber photographs of cosmic rays.* To be able to work with Blackett and take part in the study of the high-energy cosmic radiation having an unknown origin in the universe was a young man's dream. Such was my arrogance arising from the rapid successes in Tyndall's laboratory that it never for one moment occurred to me that Blackett would not immediately offer me the post.

The afternoon of that interview produced the worst of all possible results. Blackett could not decide. He intended to interview another young man from Cambridge who already had some experience with the expansion cloud chamber technique. He would write to me as soon as possible. I had expected to return to Bath, where Joyce, on vacation from her college, would be waiting for me and where I was to play cricket next day. But now I had no option but to take a train to Manchester. That was my first visit to the city. It was dark and wet when I arrived, looking exactly like the image I had built up for myself. Forty years later I was to be made an honorary freeman of the city in a splendid ceremony in the Town Hall. On that August night of 1936 I felt like a tramp waiting for a bus outside that building, and no such event could have seemed more im-probable.

The next morning I was interviewed by Bragg and his staff and immediately offered the post of assistant lecturer. In later years I still feel an element of shame at my response—to the effect that I could not give an answer, because I would prefer to work with Blackett in London. It is a measure of Bragg's stature and generosity that, although he had an alternative candidate, he did not penalize me for my impertinence and that he simply asked me to let him

---

*Blackett's collaborator in this work was Giuseppe Occhialini. A few months earlier similar evidence had been obtained independently by C. D. Anderson in the United States, and historically Anderson has the credit for the discovery of the positive electron.

know my answer in a few days. I returned to the West Country and immediately joined a cricket tour in Devonshire, out of contact with either Blackett or Bragg and unaware that Bragg, a Nobel laureate, was now demanding my answer by telegram. Eventually it was Blackett who informed Bragg that he intended to appoint J. G. Wilson from Cambridge as his assistant because he had experience with the cloud chamber technique and I did not. That is how I reluctantly agreed to be appointed an assistant lecturer on Bragg's staff in Manchester. It is an appalling story on my part, which I repeat in all humility as a warning to the young of today that there are very few who would now exhibit the kindness of Bragg and Blackett in the face of such arrogant behavior from a candidate.

After that August night in Manchester I had resolved that it would be the only night I would ever spend in the city. The story of this book is of the events that took place so that thirty years later I was rejecting all enticements to leave the university.

In the beginning there was not the slightest indication that such a bond could ever materialize. The laboratory that had such a splendid reputation derived from the heritage of the sequence of distinguished professors of physics Sir Arthur Schuster, Rutherford, and now Bragg was a dark and grimy Victorian building. Separated by forty yards there was a "new" building with the same atmosphere. I was given a room in this building—a sub-basement room with a window that faced on to a brick wall. I had never visited another physics laboratory and had innocently assumed they would be similar to the H. H. Wills Laboratory, with the same basic equipment for a wide variety of researches. The Manchester department was dominated by the X-ray crystallographic researches for which Bragg was famous. I tried to interest myself in this research but failed completely to do so. Although I had completed my Ph.D. task on thin films, I was still wrapped up in the subject and had many researches in mind on the topic.

In later years I realized this was not the only reason why I did not enter Bragg's field of research. I did not have the urge to work in one that had already been pioneered and developed. I have never been the kind of research worker who can absorb masses of existing literature on a subject and move on to new aspects of that research. One has only to reflect on the magnificent results of subsequent X-ray investigation of materials in Bragg's laboratory (molecular biology, for example) to appreciate the great opportunities for those who can proceed in that way. All research is pioneering, but I was becoming the kind of research worker who wanted a nearly virgin field to explore. That, of course, is a retrospective analysis, but it does explain why I failed to interest myself sufficiently to join Bragg's researches. This attitude to research continued throughout my whole career and eventually dominated my decision about personal participation in the researches when the telescope became operational in 1957. Instead of working on the topics already in progress, I felt an urge to begin a new subject and so undertook the investigation of the flare stars. This attitude to research runs counter to the modern trend where productivity, measured by numbers of research papers published, is a criterion of success.

Fortunately, in my youth the judgment of the value of a research project in terms of numbers of published papers and citations of these papers by other workers had not emerged, and Bragg placed no pressure on me to join his team, and so I made the bad mistake of attempting to continue my thin-film research in Manchester. No appropriate facilities existed in the laboratory. There were no vacuum pumps, no liquid air, and no glassblower. I was directed to a glassblower through several streets, but he was hopeless and had never even worked with Pyrex glass. I then pleaded with Burrow in Bristol to make the equipment I needed. He did so, but I had neither the experience nor the will to make much progress when all around me was concentrated on crystallography, and eventually I completed my thin-film researches during the Christmas vacation of 1937 in the Bristol laboratory.

The Arrival of Blackett

By that time two events had settled my future life and research career. In September 1937, a year after my appointment to the Manchester staff, I had married Joyce Chesterman. My salary was

37

£300 per annum, but it was just enough in the days when a bus ride to the laboratory cost only a few pence. So it was to Manchester that I took my bride, and the next day the telephone rang. It was a senior lecturer on Bragg's staff demanding that I go to the university to deal with examination papers. We never did have a conventional honeymoon.

A few days later I received a surprising letter from Blackett: "Hartree tells me that you may like to start on a cosmic-ray experiment." Douglas Hartree was the professor of mathematics in the university. He and his wife, Elaine, were prominent members of Manchester society and accomplished musicians. In fact, it was rumored in the university that no one would be appointed to Hartree's staff unless he could perform Bach adequately. He had been involved in my appointment, and I was soon summoned to his house for a musical evening during which I was to play a Bach prelude and fugue on their Steinway concert grand. In this way I became involved with the Hartrees, who were immensely kind when I arrived a year later with my young bride, introducing her to outstandingly friendly groups in the university and revealing to us both that the city had attributes more valuable than the superficial rain and fogs that confronted us.

Because of the nature of his mathematical interest, Hartree was closely associated with the physics department (and was later to be appointed professor of theoretical physics). In the basement of the physics building he had constructed a beautiful mechanical differential analyzer. In effect this was the mechanical analogue of the electronic computers that were developed after the war. Hartree's machine could solve complex differential and integral equations, but setting up the machine to do so was an engineering operation. In the autumn of 1936, when Douglas Hartree realized that I was not intending to become an ardent crystallographer, he suggested that I might like to help with his computing machine. I was glad to do so and spent many profitable hours with that elegant machinery. This association with Hartree in turn led to his suggestion to Blackett that I was not wholly committed to a research school in Manchester and that I might like to "start on a cosmic-ray experiment."

An extraordinary turn of events was about to lead to my long association with Blackett, all hope of which I had abandoned just a year earlier. In the spring of 1937, only six months after my arrival in Manchester, it was announced that Bragg was to leave the university to become director of the National Physical Laboratory (a few years later he succeeded Rutherford as Cavendish Professor in Cambridge). There was much speculation as to who would succeed him in Manchester, and even in November I received a letter from Appleyard quoting the Bristol gossip with a betting order of precedence on James Chadwick, Blackett, John Cockcroft, Mark Oliphant, and Edward Appleton. However, by that time in Manchester we knew it would be Blackett, and a load of equipment with which I was to begin my work on cosmic rays had indeed already arrived from Blackett's London laboratory.

After my eager response to Blackett's September letter, he asked me to come to London to take over a small automatic cloud chamber, transport it to Manchester, and put it in working order as soon as possible. Blackett was one of the great masters of experimental technique of this century. The expansion cloud chamber, in which condensation of ions would reveal the tracks of ionized particles, was the invention of C. T. R. Wilson. When he began his studies of cosmic rays, Blackett had devised a beautiful automatic system such that only when a cosmic-ray particle passed through the active region of the chamber would it be triggered and the passage of the particles photographed. In Birkbeck College, London (to which Blackett had gone from Cambridge as professor of physics in 1933), a young Chinese student, Hu Chien Shan, had constructed a small automatic cloud chamber on these principles. However, this student had decided that he must return to his homeland to help in the fight against the Japanese who had invaded his country. It was Hu's equipment that Blackett wanted me to collect and reerect in Manchester. By the time Blackett arrived to assume his post as Langworthy Professor in the physics department, at the end of 1937, I had this system working and photographing cosmic rays. Once more I had the kind of complex and delicate experimental equipment that thrilled me. For a year and a half until war wrenched me from the laboratory, I had the same kind of productive period that I had enjoyed in Bristol, and the research developments of those months were to lead to a new destiny after the end of the war. The equipment of Hu's that I had inherited, and the modifications and developments that I soon made to adapt it for particular investigations of the cosmic rays, completely fascinated me. It was like a huge magnet from which I could not remove myself, so that my absences when I should have been attending to students in the practical laboratories became so frequent that I had to be reprimanded for not fulfilling my teaching duties. The events in the cosmic rays for which I was soon searching were of rare occurrence, and the equipment was made to run throughout day and night, with only the regular change of film needing attention. A darkroom was near the equipment. I had special developing and fixing tanks to take a square frame on which I could wind yards of film around pegs in a spiral. The impatient removal of the barely fixed films and the hasty look for the photograph of the tracks for which I was searching, I remember as among the most exciting and tense moments of my whole research career.

With this equipment I immediately became involved in one of the major arguments of the day among physicists. Today a multiplicity of the fundamental particles of atomic structure are recognized. In the mid-1930s only the positive and negative electrons, the proton, the neutron, and the neutrino had been accepted as fundamental. Some years earlier the Japanese theorist Hideki Yukawa had postulated the existence of a particle about 180 times more massive than the electron. Originally known as the heavy electron or mesotron, this particle was later called the meson. The question at issue was whether the meson was a component of the penetrating cosmic radiation observed at sea level. Blackett and J. G. Wilson (whom he had employed instead of me in 1936) had made detailed studies of these phenomena and had reached the conclusion that the theory governing the interaction of the primary particles failed at high energies. An alternative view strongly held by the foremost theorists of the day was that the heavy electron (meson) was involved in these interactions and that the theory of high-energy interactions of electrons was correct.

As soon as Blackett arrived in Manchester, he told me to design an electromagnet so that I could measure the energies of the particles in my cloud chamber in order to secure more data about these phenomena. I designed the magnet to fit around my cloud chamber, journeyed to Trafford Park, and persuaded the firm of Metropolitan Vickers to make it. I placed a lead plate across the center of the cloud chamber and made an array of Geiger counters so that the expansion would be triggered only when a single particle entered the chamber from above, but a large number emerged from beneath. Then on a successful photograph I would see the track of a single particle striking the lead plate from above but a shower of particles emerging. I would then project an enlarged image of a successful frame onto a screen and measure the curvature of the tracks and thereby calculate the energies. By the time my paper describing the results of this work was published, the war had begun and I was elsewhere. However, the results were produced as part of the mass of data discussed at a memorable weekend conference in Manchester attended by many of the disputants in the argument about the existence of the meson. Werner Heisenberg, then professor of theoretical physics in Leipzig and one of the most famous theorists of the day, was there, and no one present could forget the protracted arguments and the clash of opinion between him and Blackett about this issue. By the Saturday evening there was stalemate, but when the conference reconvened on Sunday, Blackett began to concede ground. It was primarily Heisenberg who had convinced him that the theory of the interaction of high-energy electrons with matter was correct and that the heavy electrons (mesotrons or mesons) were confusing the issue because they were a prominent component of the cosmic radiation at sea level that we were studying.

At the time of this conference Europe was already in a state of tension. Large numbers of refugees came to Blackett's laboratory. Several continued their journey to America, but Blackett succored as many as possible, and those of us who were young had the immense good fortune of working among some of the finest scientists in the world, ejected from Germany and Italy at that time. Heisenberg was not one of these. He was a brief visitor who did not conceal his pro-Nazi sympathies. He stayed with the Hartrees, and after his departure Elaine Hartree told me that when saying good-bye he had handed her a rolled package. "Oh dear, he's giving me some Nazi propaganda," she thought. She unrolled it with apprehension—it was a fine copy of the autographed score of one of Beethoven's last sonatas, which he had performed excellently during his stay with them. Heisenberg remained in Germany throughout the war and worked for the Hitler government on the problem of producing an atomic bomb.

My paper on the heavy electrons in the cosmic-ray showers, "Shower Production by Penetrating Cosmic Rays," was one of only three accounts published of my researches on cosmic rays during that brief prewar period. The other two, much shorter communications published in *Nature*, concerned phenomena that, although this could not have been foreseen at that time, were to be the focus of my research for the rest of my career. For years Blackett had been concerned with the cosmological significance of cosmic rays, and his speculations on this topic, in the few articles he had asked me to read when he knew I would join his team when he came to Manchester, fascinated me. In particular, he had expressed the view that the law of the energy spectrum of the primary particles from space (that is, the way in which the number incident on the atmosphere decreased as the energy increased) might have a deep cosmic significance in a manner equivalent to the relevance of Hubble's constant for the expansion of the universe.

That was of such interest that Blackett suggested to me and J. G. Wilson (who had accompanied him from Birkbeck) that he should get another small cloud chamber and place it in coincidence with mine at different distances so that we could determine the number of particles in the extensive showers at various separations. These measurements had to cease in the summer of 1939 when Wilson and I were summoned to other tasks. At that time we had recorded extensive air showers from an estimated primary particle with an energy of 10^{19} electron volts—far exceeding anything observed previously or, indeed, for a long time thereafter.

My appointment as an assistant lecturer was for three years only and would terminate at the end of September 1939. My research in Blackett's team had gone well, but in those days there was no procedure whereby my type of appointment would lead to higher levels in the department. There was no such vacancy, and Blackett, unwilling to lose me from his team, had secured for me a senior DSIR fellowship. With this expectation of release from teaching duties it was agreed that I could take a load of cosmic-ray equipment to the high-altitude observatory at the Pic du Midi, in the Pyrenees. There I intended to explore further the high-energy spectrum of the primary particles before their energy had been dispersed among the multitude of secondaries in the lower atmosphere.

During the summer of 1939 I began to accumulate the cosmic-ray equipment and generators for the electrical supplies that I intended to mount in a van and drive across France to the Pic du Midi in the early autumn. By the end of July I had found a suitable van and was about to purchase it when I was urgently summoned to the telephone. It was Blackett in London. "Abandon all your plans for the Pic du Midi. There is another task for you." A few days later a new era in my life had begun.

5

The War Years, 1939–1941: The Development of Centimetric Radar

The increasing tensions in Europe, and the distress of the many eminent scientists fleeing from Germany and Italy as refugees who sought succor through Blackett, made a disturbing background to life in 1938 and 1939. There was no television, but the newsreels in the cinemas revealed the devastation of the bombing in Spain and the militant march of Hitler toward the west. Many of us who were young were supported by the idealism of our upbringing and for some time tended to regard these events somewhat dispassionately as disasters that affected other people but would not reach us.

Blackett was, on the contrary, a realist whose attitude and convictions were based on the assumption of inevitability. As a boy he was destined for a naval career, but this training had been interrupted by the outbreak of war in 1914. He had fought in the Battle of Jutland and had no illusions about the nature of war.

It is a remarkable fact that when Blackett came to Manchester in 1937 neither I nor any of my colleagues knew of his close association with the Air Ministry or of the cardinal part he played in preparing for the defense of the country against air attack in those years. He was frequently away from the laboratory, but the assumption was that these absences were concerned with his extensive scientific activities. We did not know until many years later that, since January 1935, he had been one of the key members of the Tizard committee, set up to advise the government on "how far recent advances in scientific and technical knowledge can be used to strengthen the present methods of defence against hostile aircraft." The work and pressure of this small committee led to the evolution of radar in the United Kingdom and to the construction of the coastal radar stations that were to play such a crucial role in the defense of the country.

Soon after the war there emerged the story of Blackett's historic association with Sir Henry Tizard in the 1930s, which led to the development of the coastal defense radar chain, and then I knew why in that phone conversation of July 1939 he was so certain that I must be prevented from setting out on my journey across France to the Pyrenees. During the war Blackett developed the science of operational research, first with the Anti-Aircraft Command of the army, then with Coastal Command of the Royal Air Force, and from 1942 with the Admiralty, where he became the director of naval operational research. Throughout those years he had been closely involved with the various committees concerned with the development of atomic weapons, but in the immediate postwar years his views diverged so much from those of the establishment that he was excluded from the main advisory circles of the government. The publication in 1948 of his *Military and Political Consequences of Atomic Energy* caused an uproar, particularly in the United States, from which he was politically excluded.

Blackett's political influence returned with the formation of the Labour administration under Prime Minister Harold Wilson in October 1964. He had been the principal architect in the development of the scientific and technological policy of the Labour party that led to the new Ministry of Technology. Prime Minister Wilson had wanted Blackett to go to the House of Lords and become the minister, but he refused and served as the deputy

45

chairman and scientific adviser to the ministry. His association with the ministry lessened when he was elected president of the Royal Society in November 1965. Blackett did eventually accept a life peerage in 1969—"The most reluctant Peer I've ever known," remarked Harold Wilson in a memorial speech after Blackett's death.

In the summer of 1939 I knew nothing of those aspects of his life concerned with the development of radar. Even at the end of July 1939, when he told me to report to Bawdsey on the east coast of England and I asked why, he merely responded, "You will soon learn." Such security and secrecy is hard to imagine in the world of the 1980s.

Hitler had occupied Austria in March 1938, and if I had known of Blackett's contacts with the inner circle of the government any lingering idealism would certainly have vanished. There is a diary note of May 1938 in which I refer to a lunch hour with Blackett: "there's a b—— smash coming, it's coming soon. Hore Belisha (secretary of state for war) expects to fight Italy and Germany before the autumn. I saw him last night. . . ."

Some time later I was talking with Blackett in the physics laboratory when an event occurred that became impressed in my memory. He had just returned from a visit to Holland and was at that time very much excited by the cosmic-ray showers I was recording simultaneously in two expansion chambers. It was early afternoon on the Saturday of 1 April, and we were walking up the stairs to his office from the laboratory where my equipment was working. As we reached the halfway landing, a telegraph boy appeared at the foot of the stairs and handed Blackett a cable. He tore open the envelope, read the message, and handed it to me. It was from Victor Weisskopf in America urging Blackett to use his influence to prevent the publication of some recent work on the fission of the uranium nucleus, because of possible misuse in Europe. When writing this chapter forty-eight years later, I asked Weisskopf if he could remember the circumstances that led to the dispatch of this cable. He told me that during a visit to Princeton he found Leo Szilard, Eugene Wigner, and George Placzek agitated because they feared that the Nazis might develop an atomic bomb if they were aware of the results of Frédéric Joliot and his co-workers in Paris, who had measured the number of neutrons released in the fission of the uranium nucleus. One of Joliot's co-workers was Hans von Halban, a close friend of Weisskopf's, and it was to him and Blackett that Weisskopf sent the cable. I asked Blackett what he would do. He replied that he could do nothing. Neither did the request to Joliot have any effect, and he replied that he saw no reason for holding back the physical facts he had discovered. I never knew whether Blackett had any vision of what this discovery meant for the future. I certainly did not and thought in terms only of subatomic energy at last.

The period of under two years that I had worked with Blackett was merely the beginning of a long association with him that was to last through war and peace for another forty years and have consequences that neither of us could possibly have foreseen. When he came to Manchester, his brilliance and inexhaustible energy transformed the grimy Victorian laboratory into a physical and intellectual palace to which many of the great scientists of the world migrated both before and after the war. He was ruthless in his opinion of those who could not compete with his arguments and practical skills. I often wondered why I survived in his favor, since he and his wife, Pat, were so cosmopolitan that to be an ordinary English youth was not a good beginning. English was often a rare language to be heard at their parties. However, I could make Geiger counters where others were afraid of the blowtorch, and although I could never approach his extraordinary experimental gifts, I adored complex equipment, made it work, and produced results. Above all, by the time he came to Manchester, I had married the girl who charmed everyone with whom she came into contact, and Blackett was no exception. It was a relationship epitomized by a telephone call from his wife to Joyce early one evening not long after our first child, Susan, was born. "We want you to come here to meet some visitors tonight. If you can't get anyone to mind the baby at this short notice, then although we'd like to see you both we would rather have *you* if only one of you can come." Toward the end of his life, when he was aged but still handsome and there was time for sentiment, Joyce was on his arm leaving a banquet in London. He paid her some compliment, and she replied flippantly, "You know, Patrick, you're the only person I would ever have left Bernard for"—"Why didn't you?"—"Because you never asked me." I am glad he did not—otherwise the story that follows would never have occurred.

The Outbreak of War

It was 29 July when Blackett advised me to abandon all the plans for purchasing the van to drive the cosmic-ray equipment to the Pic du Midi, and two weeks later I had my first introduction to a new kind of life. By then peace and peacetime activities had become so uneasy and unreal that it seemed only a question of when we should be plunged into war. I was not alone among the young people who had grown to despair of the type of peace of that summer of 1939 and had developed an eagerness to begin whatever tasks might be assigned to us in the confrontation with the Fascist dictators of Europe. I had not long to wait. On 4 August I switched off my expansion chamber, covered the equipment with dust sheets, and said good-bye to the laboratory. On the fourteenth I was ferried in an Air Ministry launch across the river to Bawdsey Manor, on the east coast near the town of Felixstowe. There was a cricket bag in the hall, but if I imagined that gave any guide to the life ahead, I was quickly disillusioned. A. P. Rowe, in charge of the establishment, unfolded to us the secrets of radar and of the air defense of the country, and for two days we toured the laboratories and the operational radar station, part of the CH (Chain Home) network guarding the coasts of the country. I was astonished at the magnitude of the equipment, using techniques and practices far beyond those of my university experience. A few days later we were at the Fighter Command headquarters in Stanmore, Middlesex. In the spacious underground, bombproof building we saw the filter room and a demonstration of the radar echoes from an "invading" French squadron and the girls in air force uniform placing the symbols of the "enemy" aircraft on a vast map of England so that the commander in chief could transmit the directions for interception to his fighter squadrons. Then we went to one of these squadrons at Biggin Hill and my first sight of the interior of a Hurricane fighter and of the scrambling of the squadron to intercept the "enemy" aircraft. In those few days an entirely new world had opened before my eyes, impressive and dignified. It was to be my world for the next six years.

After this initiation I was sent, nominally in charge of a small group of university scientists, to one of the operational coastal defense (CH) radars at Staxton Wold, near Scarborough, in Yorkshire, in order to gain further acquaintance with the radar equipment and its operation. Fate had decreed that I would be in the receiving section of that massive installation listening to the broadcast of Prime Minister Chamberlain on the Sunday morning of 3 September. Germany had invaded Poland on 1 September, and our ultimatum to Hitler had expired at eleven on this Sunday morning. We were at war with Germany, and at that moment I saw the radar echoes that were to form my future destiny.

Scone Airport

Before the end of that month I was given an Air Ministry appointment as a junior scientific officer. The main Air Ministry Research Establishment at Bawdsey had transferred to Dundee, in Scotland, fearing that the exposure of Bawdsey Manor on the east coast of England would make it an easy target for the German bombers. The city of Perth lies about thirty miles west of Dundee, and the small settlement of Scone was a few miles from that city. The famous stone of Scone was the seat on which the kings of Scotland had been crowned until the Act of Union with England, early in the eighteenth century. E. G. ("Taffy") Bowen had taken his airborne radar section of the establishment to a small airport at Scone, and it was to him that I reported.

Bowen was one of the few early pioneers who developed the embryonic radar systems after the demonstration early in 1935, which showed that it was possible to detect radio waves reflected from an airplane when it passed through the beam from the transmitting station. Bowen was also one of only four scientists who assembled transmitting and receiving equipment on a remote long spit of salt marsh and shingle at Orfordness, on the east coast about ninety miles northeast of London. Their equipment worked on a wavelength of 50 meters, and by mid-June 1935 they had successfully demonstrated the potential of the system for the detection of aircraft. Interference from commercial radio traffic soon forced them to reduce the wavelength, and it was in the wavelength range of 10 to 13 meters that, before the end of that year, the vital step was taken to build the massive radar stations (CH) that a few years later played their historic role in the Battle of Britain.

In the spring of 1936 the Air Ministry purchased the beautiful Manor of Bawdsey, lying at the confluence of the river Deben and the sea. On a small hill in that estate the 240-foot-high towers of the first radar station soon appeared, and it was there in August of 1939 that I had my first introduction to radar. This radar and the similar ones that had been built along the east and south coasts of England were equipped with specially designed powerful transmitters. They generated pulses of radio waves at the rate of fifty per second. The power in the pulses was several hundred kilowatts; the transmitter valves were continuously evacuated and water cooled. These radar pulses, each of about twenty microseconds' duration, were transmitted through the aerials mounted on the high towers so that the radar beam would cover the approaches to the English coast. The succession of such stations, spaced at about thirty-mile intervals, was arranged so that the complete radar screen covered the approaches to the east and south coasts out to ranges of more than a hundred miles. Any aircraft flying into this beam would scatter some of the energy back to the receiving stations, where it would appear as a "blip," or radar echo, on the cathode-ray tube from which the range and approximate altitude of the aircraft could be measured.

During the summer months of 1940, when the Germans entered on their massive bombing campaign of England, the few fighter squadrons of the Royal Air Force could be retained on the ground until the radars revealed that a raid was approaching. Sufficient information could be given by the radars so that the operational commanders could direct the fighter squadrons to within visual distance of the enemy aircraft. This ability to keep the wear and tear on the fighter aircraft and pilots to a minimum became a vital factor in the Battle of Britain, so that by the autumn of 1940 the German Luftwaffe was forced to abandon the daylight bombing raids.

It is a remarkable fact that two years before the outbreak of the war, as soon as Sir Henry Tizard and his committee were satisfied that the ground-based radar chain would enable our fighters to be placed in contact with enemy aircraft during daylight, Tizard had foreseen that the Luftwaffe, frustrated by day, would turn to night bombing. Bowen had been directed away from his work on the massive ground stations with instructions to develop a miniaturized version of radar to be carried in the night-fighter aircraft.

The interception of enemy aircraft in darkness was a formidable problem. The range at which a pilot could identify an enemy aircraft at night was less than 1,000 feet, and the ground-based radar systems could not possibly position a night fighter with that precision. The solution was to develop a miniature radar small enough to be installed in night fighters so that the pilot could detect the enemy aircraft at a range of several miles and close down to the range of 500 to 1,000 feet, where it could then be seen and identified visually.

The techniques of miniaturization in those days did not exist, but it was not simply a case of reducing the size of the ground stations by thousands of times. The night-fighter radar had to be entirely self-contained, and from the equipment carried on the aircraft the pilot had not merely to locate an enemy aircraft in his vicinity but also to be able to decide in which direction to fly to make the interception.

From the archives now accessible it is evident that Tizard and the members of his committee appreciated in 1937, when the need for night interception was first seen, that the essence of the problem was to find some means of generating radar waves on wavelengths much shorter than that used on the ground stations. The nature and shape of the beam transmitted (or received—the two are identical) by an aerial is referred to as the polar diagram. The width of this polar diagram depends on the aperture of the aerial and on the wavelength. For a given wavelength the larger the aerial, the narrower the polar diagram, and for a given size of aerial the polar diagram narrows as the wavelength is reduced. Since only very simple and small-sized aerials could be fitted on night fighters, it was desirable to find means of reducing the wavelength so that a narrow beam could be transmitted from the fighter.

Before the war began, Tizard was raising this question with a number of his close associates and with the firms already involved in the manufacture of the ground radar systems. In retrospect it seems remarkable that even at that stage he was stressing the importance of finding some means of generating significant powers on wavelengths of a few centimeters instead of the several meters then in use. Eventually the brilliant solution evolved in the physics department in the University of Birmingham, where John T. Randall and his assistant H. A. Boot invented the cavity magnetron—essentially a solid block of copper with accurately machined cavities around which electrons would circulate in a magnetic field. This device could generate large amounts of power in the centimeter wave band and was destined to have a vital effect on the Allied war effort in radar systems far removed from that of air interception, which was the primary stimulus for its development. In these matters I was to become deeply involved, but that lay in the future since the first magnetron was not available until the summer of 1940.

By the time of my visit to Bawdsey in August 1939 Bowen and his group had successfully built a small airborne radar system working on a wavelength of 1.5 meters, using specially developed but conventional thermionic valves, and had demonstrated the potential for air interception and for the detection of surface vessels on the sea. The code names AI (Air Interception) and ASV (Air to Surface Vessel) came into general use for these airborne radar systems.

Suddenly, with war imminent, the Air Ministry placed orders for thirty of these airborne systems with a demand that they be fitted into thirty squadron aircraft by 1 September. There was no engineered version of the equipment, and such demands were unrealistic—particularly since separate firms were involved in the manufacture of the transmitters and the receivers. It was into this muddle and chaos, with Taffy Bowen's group attempting to fit the manufactured version of an experimental system into operational aircraft, that I was plunged when I arrived at Scone airport late in September 1939. I was the first of the university scientists to be placed in Bowen's group. Others were to follow, and in his own account of this time Bowen wrote of us, "It took them some time to get used to the rituals of government service, and this was not made any easier by the survival conditions in which we found ourselves. Moreover, they were not always allocated the most glamorous of jobs." Indeed, this new life came as a shock to me. The aircraft hangar in which we worked, the lack of equipment, and the general disorganization seemed ill matched to the urgent operational needs for the equipment passing through our hands. Bowen later remarked, "It was a time when our resources were strained to the limit and we were under tremendous pressure to get AI and ASV radar into fighting aircraft. . . ." Of us he recalled that at least we "were fresh and not suffering from the physical and mental strain which the pressures of the previous few months had brought to many of us."*

Indeed, I was fresh and found the muddles and indecision so ill fitted to the defense priority that was attached to our task that I soon wrote a long and bitter letter of complaint to Blackett. His response was a reprimand that I was to remember for the rest of my career in war and peace:

> You underestimate the difficulties of getting dispersal organizations going properly and of competing with the general expansion. You must be more tolerant too, especially at first till you have really achieved something in these technical fields—then you can criticize safely. . . . One of the great things to remember is that all defence work like all service work is very much a matter of dealing with people—the qualities of the personnel are part of the experimental facts and it is no use getting too upset about them.

The Problems with the AI Radar

When I first met Bowen in his small office inside the hangar at Scone airport, he explained that his main concern was to get an

---

*E. G. Bowen, *Radar Days* (Bristol: Adam Hilger, 1987), p. 90.

elemental AI (Airborne Interception) system operational, so that our night fighters could locate and maneuver within visual distance of the enemy bombers. When I became acquainted with this system late in September 1939, the equipment was fitted in Blenheim aircraft. The wavelength of 1.5 meters was determined by the availability of vacuum tubes capable of generating any significant power on short wavelengths at that time. Even at this wavelength any conceivable size of aerial still produced a broad beam, and Bowen had developed a system involving simple dipole aerials on the wings and fuselage in an attempt to overcome this deficiency. A navigator was given a cathode-ray tube display and, by comparing the strength of the radar echo from the target aircraft received on each of the dipoles, could give the pilot instructions to turn left or right or to increase or decrease his altitude. Although quite a triumph of miniaturization for that time, the system was barely operational. There were two fundamental defects. The maximum range was limited to the height above ground at which the aircraft was flying, because ground returns obscured echoes from an aircraft at greater range—and that meant about 10,000 feet in those days. This limitation could be overcome to some extent if the fighter worked closely with a ground station so that the pilot could be directed to within about two miles of the target—and indeed no significant success was achieved in the night air battles until such a ground-controlled interception system became operational, in the winter of 1940–41. The second defect arose from the crude nature of the transmitting and receiving system, which limited the minimum range to about twice the range at which the pilot or gunner could obtain a visual silhouette of the bomber on a dark night. It is hardly surprising that by the end of October 1940 only one enemy bomber had been destroyed with the aid of AI.

The urgency of combating the night bombers was such that many avenues were being explored to overcome the deficiencies of this equipment. Among others, the firm of EMI (Electrical and Musical Industries) had been given a development contract, and A. D. Blumlein (with whom I was later associated under tragic circumstances) introduced substantial improvements to the electronic system. The improved system (AI Mark IV), with more power and better aerials, gave a maximum range of about 20,000 feet and a minimum range of around 600 feet. Installed in Beaufighters, this system combined with ground control contributed greatly to the defeat of the German night bombing during the winter of 1940–41.

By that time I had very little connection with the 1.5-meter AI system, since Bowen had soon directed my work to the ultimate solution for AI—that is, the use of much shorter wavelengths so that a feasible aerial system producing a narrow beam could be built onto the fighter. The vacuum tube designers (and particularly the General Electric Company of London) were already at work on the problem of producing substantial power at wavelengths of 50 centimeters and shorter. Within a few days of my arrival at Scone airport, I had been told to get a 50-centimeter system working. The workshops and the library were miles away in Dundee, I knew very little of these techniques, and my dismay was increased a few days later when at the beginning of October I was told that the immediate operational pressures were so severe that I must give my attention to improving the 1.5-meter system fitted in the Blenheims. Soon it became apparent that whether I spent the day on the 1.5-meter or on the 50-centimeter depended on whether the opinion of Bowen or of Gerald Touch (his deputy) held sway. Bowen looked to the future and realized that the shorter-wavelength equipment must be developed, whereas Touch's constant refrain to me was that there was a "war on" and that I should give my attention to the apparatus that already existed in the aircraft.

In the jet engine age of the postwar world there is so much organized sameness that it may be hard for an individual to remember, or even to experience, the thrill of a first flight. My first flight in an airplane was not of that character. On a Sunday morning early in October I was fitted with a flying suit and parachute. Sitting on a parachute on an iron bar next to the pilot of a Blenheim night fighter that bumped its way across the grass field for takeoff was an introduction to wartime flying that I was unlikely to forget. In the absence of any intercom the engine noise made conversation impossible, and the radar echoes on the AI equipment from the target aircraft seemed far less interesting than

55

the sight from 10,000 feet of the snowcapped Scottish mountains, the city of Perth, and the river Tay scintillating in the brilliant sunshine. However, the landscape over which I flew with this early airborne radar was soon to take on a very different aspect— our stay at Scone airport was short, and in the first days of November we moved to South Wales.

St. Athan

St. Athan, some miles west of the small port of Barry, in South Wales, was an enormous unfinished Royal Air Force airfield then being established primarily for airframe and engine maintenance. Now it was given the additional task of installing airborne radar, and Bowen's group, a couple of dozen strong, was sent there to instruct the unit on the fitting and testing of airborne radar. This was intended as a temporary assignment for Bowen's group because of the urgency of producing operational AI- and ASV-equipped aircraft. It was very far from my vision of the research and development tasks that I had anticipated, and the hangar in which we were housed bore no resemblance to any laboratory I had ever contemplated. The improvised laboratory was one of the open annexes around the walls of the hangar through which the wintry blasts were already blowing. We were given some canvas partitions to protect one of the annexes from the vast hangar space. Nearly half a century later Bowen recalled his memory of me in this annex:

> We might have got by with those canvas partitions if it were summer, but it was already November and for several months there was no source of heat whatsoever. For a large part of the time our people wore great-coats and gloves inside this improvised laboratory space. The sight of a future University Professor, in coat, gloves and cap, working on a radar chassis with a soldering iron will stay with me for the rest of my life. The conditions were simply appalling—and any member of the airborne group will say that this was by far the most unpleasant and least pro-ductive part of their lives. Here was one of the most sophisticated of

defence developments being introduced to the Royal Air Force, and it was being done under conditions which would have produced a riot in a prison farm.*

This move exacerbated the seesaw between the present and the future. Early in December we received an urgent signal from the Air Ministry to cease working on AI in order to concentrate on fitting the 1.5-meter equipment in Hudson aircraft for Coastal Command. Just before Christmas we tested the first Hudsons equipped with the 1.5-meter ASV over the Bristol Channel, and my diary entry for 21 December mentions that the ASV worked magnificently. Alas, I was soon engulfed in the first of the two wartime tragedies that surrounded my work. On Sunday, 7 January 1940, the two young men who had come with me from Manchester were testing another Hudson with the ASV equipment when it crashed into a mountain near Bridgend. Peter Ingleby had worked with me on cosmic rays, he was my roommate, and he and R. K. Beattie had worked with me at Scone airport. They were my only helpers and my only contact with a world that seemed then to have disappeared forever.

Patrick Blackett soon roused me from these depths of despair. When he heard about the crash, he immediately came to St. Athan and inspired a hope for the future. This was not his first visit to St. Athan. In November he and Sir Henry Tizard had come to see what we were doing. This time he was appalled at my story of our diversion to the 1.5-meter AI and then to the ASV, and his promise to restore me to research on the 50-centimeter system was soon implemented. Before the end of the month I was again concentrating on those problems. He had promised also to find a fellow spirit from Farnborough (where Blackett was working on the Mark 14 bombsight) to work with me, and on 26 February, as the snow and ice of that fierce winter began to disappear, A. L. Hodgkin appeared at St. Athan. Alan Hodgkin was one of the young men from the universities referred to in the introduction who were to have a brilliant postwar career. Then, as Blackett had promised, he was a fellow spirit, and together as we investigated possible aerials that could be used in aircraft on

*Bowen, *Radar Days,* p. 93.

these short wavelengths, we could talk and plan the future, and the wartime world of research and development at last began to emerge.

Worth Matravers

The abrupt move of the main research establishment from Bawdsey Manor to Dundee on the outbreak of war was a panic decision that bore little relation to its function. The plans to create a more suitable environment for the establishment bore fruit in early May 1940 when it was transferred to a site on the south coast of England. Some weeks earlier the dilemma about the function of Bowen's group—that is, whether it should concentrate on the immediate operational needs of the night fighters or develop the shorter-wavelength systems—had been settled. The group was split, and the majority either transferred to the Royal Aircraft Establishment at Farnborough or to the airfields to deal with the day-to-day operational problems.

Mercifully, with Alan Hodgkin I was ordered to report to the main establishment now transferring to the Dorset coast. Although these final decisions were not promulgated until 18 April, the arrangements for Hodgkin and me to take our centimeter work to Worth Matravers had effectively been decided some weeks earlier. I asked the DCD (director of communications development) where our airfield was to be, because access to experimental aircraft was essential. His response was that he understood there was an airfield in Christchurch and asked me to investigate. We discovered this to be a rather small field suitable for light aircraft of the "Gipsy Moth" type, a few of which were in wooden huts on the perimeter. When I reported this to the DCD by phone, he seemed most displeased and said arrangements would be made for a more suitable airfield, but for the next eighteen months, until the large airfield at Hurn was available, we were to make many bumpy takeoffs and landings over that field, which was just large enough for a lightly loaded Blenheim to use.

Finally on 4 May we evacuated St. Athan. As Germany invaded Holland and Belgium and Churchill succeeded Chamberlain as prime minister, we moved our centimeter equipment into a wooden hut about halfway between the main site and the cliffs at Worth Matravers. It was a magnificent situation. The blue sky and the sparkling sea created an environment that seemed exotic after the drabness of St. Athan. The small seaside town of Swanage lay only a few miles from Worth Matravers, and soon, like many others in the establishment, I found a pleasant house to rent and established a wartime home there with Joyce and Susan, now two years old. For two years we lived there—a period of the war when neither the impact of rationing nor the occasional bomb and the crack of the German machine guns greatly disturbed our life.

In that spring of 1940, when such desperate events were unfolding across the English Channel, many new people from the universities joined Rowe's establishment, and once more the corporate life emerged against which one could work hard and ceaselessly. H. W. B. Skinner, whose student I had been a few years earlier at Bristol, arrived, and on 14 May P. I. Dee (subsequently professor of natural philosophy in the University of Glasgow) came from Cambridge. Skinner and Dee were the seniors in a rapidly accumulating group of university people. By the end of the month I was able to generate sufficient power on a wavelength of 11 centimeters to measure polar diagrams and to investigate the effects of transmission through various materials that might be used for the nose of the aircraft. By 9 June we had a good enough beam of radiation from a horn to measure the reflection from a tin sheet and found that it returned a strong signal.

Forty years later all of this seems obvious and elementary, but at that stage we were entering a wavelength range in which the properties of the radiation were unknown. Of course, there were endless visitors to our site (known as C Site), and as the pressure for an airborne system on this wavelength grew, many of the senior RAF visitors became concerned about the problem of flying one of my large centimeter antenna in a fighter plane. Our concern was to produce a directive antenna system in order that the beam of radiation transmitted from the fighter would be as sharp as possible. With the experimental system on the ground we had succeeded in producing a narrow beam of radiation by using a horn-type antenna. But this type of horn antenna was over 2 meters long and we had to explore other possibilities for an antenna that might be more acceptable in a fighter aircraft. In mid-June, Nevill Mott and Charles Ellis came from Bristol with an antenna made of a wire grid in front of a flat sheet of metal for me to test. The shape of the transmitted beam of radiation was excellent, but by that time Herbert Skinner had persuaded me to make a parabola, and this, at only a tenth of the depth of the 2-meter horn, seemed to point to the solution. On 12 June the discovery that I could shift the beam 8 degrees by simply moving the dipole 5 centimeters from the focus led me to comment in the diary that I regarded the aerial problem as 75 percent solved. A few days later I had a full paraboloid under test, and by the end of June the major problem was to get sufficient power to detect aircraft at a range of a few miles. When the war was over, I was to become fully acquainted with the use of paraboloids in quite a different context. Then, access to libraries and the prewar literature indicated that the virtue of paraboloids for use as transmitting and receiving aerials on short wavelengths had been known elsewhere. It is an interesting example of the fact that scientists working in isolation in a new field frequently arrive at identical solutions to a problem.

On 19 July the first of the Randall/Boot cavity magnetrons arrived from GEC. Early in August, with this powerful generator fed into the paraboloids, we obtained echoes from the coast guard hut a quarter of a mile distant. On 12 August, W. E. Burcham, one of the young Cambridge physicists who had joined us, fixed new cables to feed the magnetron into a double paraboloid system that I had mounted on a swivel. The receiver was a simple crystal mixer and at 6 P.M. on that day we obtained what was probably the first 10-centimeter radar echo from an aircraft in flight. The next day the experiment was repeated and demonstrated with great delight to Watson Watt and Rowe.* In the evening of that day we asked one of our assistants to ride his bicycle

*On 26 February 1935 R. A. Watson Watt, then Superintendent of the Radio Research Board at Slough, arranged the demonstration that initiated the development of radar. In March 1936 he was appointed Superintendent of the newly formed radar research group at Bawdsey Manor. In May 1938 he was made director of communications development and was succeeded at Bawdsey by A. P. Rowe.

along the cliff and carry a tin sheet. He was a few hundred yards from the equipment, and the slope rising to the face of the cliff was behind him. To our astonishment a strong echo was received, which, as I wrote in the diary, "was amazing considering it should be right in the ground returns." A somewhat casual experiment and observation on that day was to have immense repercussions, which at the end of 1941 were to lead Rowe to direct me to my major wartime occupation—the development of a blind-bombing system.

However, on 13 August 1940 that moment was still sixteen months distant—a period during which our cumbersome equipment was to be transformed into an airborne system of great elegance. The maximum size of a paraboloid that could be used in the nose of the Blenheim was about two feet. The coverage provided by the beam of a few degrees from a fixed paraboloid was, of course, useless for the pilot, and throughout the late summer and autumn of 1940 tempestuous arguments occurred as to the means by which we could scan the beam over at least the forward 90 degrees and preferably the full 180 degrees. One day in August of 1940 I was driving to the airfield at Christchurch with Alan Hodgkin, talking about this problem, when he suddenly had the inspired solution—to rotate the paraboloid around the dipole in a spiral. He said he could also see how to make the spiral motion, by rotating the paraboloid at high speed around the direction of flight and give this axis a slow spiral motion.

The first airborne tests with this remarkable system (known as AI Mark VII), fitted in a Blenheim, were made on 10 March 1941. In retrospect the transformation of the ungainly ground system that had first detected an aircraft only seven months earlier seems miraculous. By June of 1941 ranges of six miles were being obtained on target aircraft, and a development program for the fitting of Beaufighter aircraft, which were superseding the Blenheims for night fighters, was proceeding. In April 1942 a Beaufighter achieved the first destruction of a German night bomber by means of this centimeter equipment. In the subsequent spring nights of 1942 this AI enabled our night fighters to inflict heavy casualties on the German night bombers.

After the first tests of the centimeter AI with Alan Hodgkin's spiral scanner, the major problem of transferring the experimental system to a manufactured version and the fitting of operational aircraft did not actively involve me, and like others in Dee's group, I started work on another project. I began to develop an AI system that would "lock on" to the target aircraft and enable the pilot to fire blind without the need for visual identification.

This was given the code name of AIF, and the development of an experimental system in the grounds of the school at Leeson House, which looked over Swanage Bay, had gone very well. I had developed a split-beam technique using an offset dipole rotating at 600 revolutions per minute, and in collaboration with Hodgkin, F. C. Williams (one of the pioneers of the postwar computer developments), and Metropolitan Vickers, on the control circuits and scanners, I had soon produced a ground system that followed aircraft with an accuracy of better than a degree.

There was widespread interest in this system, and we were demonstrating it to many people. The value of this centimeter equipment for other operational uses soon became apparent. During the autumn of 1940, for example, I was ordered to stop developing the equipment for airborne use and concentrate on making it suitable for use by the army as a GL (gun-laying) system. In October 1940 I wrote in the diary that, after a demonstration to an American technical mission and the Inspector General of the Royal Air Force Sir Edgar Ludlow-Hewitt, Rowe told me that a gun had to be fired using the device in two weeks' time. Eventually, in December, D. M. Robinson (who later went to the USA) took over responsibility for this GL system and transferred to Rugby for development and manufacture.

At that time, too, I had interesting associations with naval problems. At the beginning of November 1940 two naval commanders came to look at the equipment. They were soon followed by the head of the Admiralty Signal School, and during the next few weeks we fitted equipment into a naval trailer that had been brought to Peverill Point, above the pier at Swanage. The tests on the detection of a small launch (the *Titlark*) were so successful that a submarine was shortly brought into the bay. These demonstrations of the use of centimeter radar so impressed the naval personnel that their own developments were soon promulgated and—at least for a time—I was allowed to return to the airborne lock-follow problem.

By the end of 1940 most of those interesting diversions were over, and during 1941 I prepared an airborne version of the lock-follow radar. By October 1941 I was able to make the first test flights with this lock-follow system fitted in a Blenheim aircraft.

One problem that disturbed the accuracy of the lock-follow was a high-speed fading of the echo from a target aircraft, and in order to establish the parameters of the system it was essential for us to have data on the pulse-to-pulse amplitudes. Someone thought he knew a young man who had a suitable photographic technique. So it was that on 18 July I met Andrew Huxley, who to my astonishment arrived with only a Leica camera. However, this turned out to be brilliantly successful. He had modified the inside of the camera so that, simply by grasping the end of the film, he could pull it across the open shutter at the speed necessary to separate the images of the successive echoes. We soon acquired all the necessary data on the pulse-to-pulse fading.

My expectation that I would in 1942 see this blind-flying lock-follow AI system through to its use against the German night bombers in the new Beaufighter night-fighter aircraft was soon to be abruptly quashed. Unexpectedly and without warning before 1942 dawned, I was summoned to Rowe's office and ordered to hand over all my responsibility for that equipment to other colleagues.

6

The War Years, 1942–1945: The Whirlwind Years

The Blind-Bombing Equipment—The Summons

For a few days around Christmas of 1941 I escaped from work. On Christmas Eve we drove to Bath and spent a happy time staying with Joyce's parents. The horror of the wartime restrictions that we were later to remember vividly had not yet gripped us. Within a few months the city of Bath and the house where we spent that Christmas were to be destroyed by German bombs, but of those forthcoming events we were blissfully unaware. Neither did I realize that within days a dramatic change was to be forced on my own wartime activities.

All too soon the brief leave ended, and when I returned to my equipment at Leeson House on 29 December, Rowe summoned me to his office and ordered me to take charge of the development of a new centimetric airborne device to help the Royal Air Force with its night bombing. The code name H_2S was soon to be allocated to this device. Eventually there was to be a variety of opinion about the origin of that code. In fact, for a short time my files were labeled BN, for blind navigation, but soon Rowe received a direct instruction from Lord Cherwell,* the prime minister's close aide and adviser on scientific matters, that it must be called H$_2$S, the chemical formula for hydrogen sulfide, because it was a "stinking" matter that it had not been devised earlier.

One positive element in the instruction was to remove any idea that the system was for navigating. My orders were that the device was to be developed for use in the heavy four-engine bombers then just coming into operational use for blind bombing, and any navigational facility was a secondary consideration. In any event, at that moment the name of the device was the very least of my worries, and I responded that I did not want to do this. Since the move to Worth Matravers I had at last been involved in satisfying activities. Apart from the scientific interest of these new centimeter radar developments, enough of the idealism of the prewar days remained to make me feel glad that I was involved in devices that would help protect the country from the German bombers and that I was not involved in destructive activities. Those last shreds of misplaced idealism were soon to be blown out of my being by a whirlwind of activities that made the previous two years of my wartime work seem like a rest cure.

At the time of Rowe's demand I had just had the interesting liaison with the naval activities and was looking forward to the possibility of getting the Beaufighters equipped with lock-follow AI radar (AIF) in service, so it was perhaps hardly surprising that I did not want to sever my connection with these activities—and that was my response to Rowe's request. On that day Dee, my immediate superior, was absent, but on 30 December he exerted strong pressure, in the course of which he drove me to Hurn, hoping that the new airfield from which we were to operate would be an enticement. On 1 January 1942 I was again taken to Rowe, who in the presence of Lewis and Dee said there was no alternative.

In his book *One Story of Radar* Rowe simply records, "On 1 January 1942 Lovell was given charge of the H$_2$S project and of the sister project, centimetre ASV." But my diary records that it was

*In 1942 F. A. Lindemann, on being raised to the peerage, became Lord Cherwell.

"absolutely heartbreaking to leave AIF" and that F. C. Williams and Ritson (a young assistant) were "terribly fed up." I added, "Still no doubt I'll recover in a few days."

The Background to H_2S

Throughout 1941 there had been an increasing recognition that our own night bombing of Germany was doing very little damage, because the pilots could not find their targets. After the evacuation from Dunkirk our main offensive became the bombing of Germany. In the summer of 1941, more than a year after the evacuation of the British Expeditionary Force, some rumors began to reach the Telecommunications Research Establishment (TRE) that our bombing offensive was proving ineffective. It was said to be unusual for a bombing sortie to reach even the target area, and the estimate was that only 10 percent succeeded in doing so. The raids were therefore little more than propagandist efforts and had no measurable deleterious effect on the German people or their war effort. Churchill later stated that he authorized Lindemann to make this investigation and that "we learnt that although Bomber Command believed they had found the target two-thirds of the crews actually failed to strike within five miles of it."* On 3 September 1941 Churchill minuted the chief of air staff and demanded his urgent attention to the problem.

There had been no appeals to TRE from Bomber Command for radar aids such as those emanating from Fighter and Coastal Command. It is true that the navigational aid known as Gee was under development, but at that time in 1941 the operational trials had only just commenced. In any event, Gee was a navigational and not a bombing aid, with a range limited to about 300 miles. The incentive was injected into TRE by Cherwell. Shortly after Churchill had sent his report to the chief of air staff, he appealed to Rowe for help, and late in October one of Rowe's Sunday meetings generally attended by high-ranking RAF and govern-

*Winston Churchill, *The Second World War*, vol. 4, *The Hinge of Fate* (London: Cassell, 1951), p. 250.

ment officials anxious to escape the bombing of London, and known as Sunday Soviets, discussed the problem of helping Bomber Command attack unseen targets. Cherwell insisted on the need for great ranges of operation, and this ruled out of the discussions the navigational aid Gee and the blind-bombing OBOE system (also in early development), which depended on ground transmissions from Britain. No one had any good ideas on that Sunday. The ill-defined reflections from the ground (known as ground returns) from airborne radar had been the bugbear of all the AI systems, but a few days later Dee realized that on our centimeter equipment we were getting echoes from the town of Swanage, lying a few miles distant down the hill from our field site at Leeson House, and he also recalled the echoes from the assistant carrying the tin sheet on his bicycle at Worth. Suddenly it appeared that the narrow-beam centimeter radars might give some information in these ground returns quite distinct from the amorphous clutter of ground echoes observed with the longer-wavelength airborne radars.

By this time Dee was formally in charge of those of us working on the centimeter developments, and it was he who persuaded B. J. O'Kane and G. S. Hensby to make a test flight with the centimeter AI equipment in Blenheim V-6000. The use of Hodgkin's spiral scanner was not the only avenue explored for scanning the centimeter beam. There were many who doubted whether this was a realistic mechanical project and so a more conventional scanning antenna had been built and fitted in this Blenheim aircraft. This antenna was a paraboloid, 28 inches in diameter, also mounted in the nose of the aircraft but rotating continuously about a vertical axis. At every rotation the elevation changed slightly so that the narrow beam of radiation swept out a helical pattern ahead of the fighter. In early November, O'Kane and Hensby modified the vertical motion of this helical device so that the scanner rotated continuously at a fixed depression of about 10 degrees to the horizontal. The beam width was 15 degrees, and the nose structure limited the coverage to about 60 degrees each side of the direction of flight.

On the Saturday following Rowe's Sunday meeting, O'Kane and Hensby flew this aircraft toward Southampton at a height of a few thousand feet and saw echoes from the town on the cathode-ray tube clearly distinguished from the general ground returns found with the meter radar. In the days following they detected many towns in the Midlands when flying above clouds, and the military installations on Salisbury Plain. After the first flight they carried a camera to photograph the cathode-ray tube. When Dee saw the prints, he seized them and rushed them to Rowe's desk while they were still wet. "This is the turning point of the war," Rowe exclaimed. However, although these records caused great excitement in TRE, their precise value to a bomber pilot was not immediately apparent. There was no question but that large towns gave discrete echo returns visible on the cathode-ray tube, but so did many other objects as well. These results soon reached Cherwell, and on 23 December 1941 the secretary of state for air (Sir Archibald Sinclair) summoned a meeting to decide what should be done. After hearing the facts, he gave instructions that six specific flights should be made "to determine whether the signals obtained in separate flights could be definitely associated with specific ground objects."

It is indicative of the temper of those times that while O'Kane and Hensby were making these six test flights in Blenheim V-6000, the development of the system proceeded as a matter of high priority on the assumption that these results would be favorable, and by the time their report was ready (23 April 1942), a Halifax bomber equipped with the first prototype H_2S system (as it had by then been coded) had made a test flight and I had rapidly become completely involved.

1942: The Early Development of H_2S

The flights of O'Kane and Hensby in Blenheim V-6000 had been made with the modified AI equipment, which gave only the forward sector on a range-azimuth presentation. It seemed obvious to us that any operational system would have to be all-around looking with a plan position indicator, and our idea was to fix a rotating scanner underneath the fuselage of the aircraft. On 4 January I traveled to the Handley Page works and faced Mr. Handley Page and his designers with our requirement for a large perspex cupola (8 feet long, 4 feet wide, and 18 inches deep) underneath the bomber. It is understandable that he raised the strongest objections, claiming that the performance of his aircraft would be ruined and the bomb load seriously reduced. The obvious reply was that it was better to drop some bombs in the right place than to scatter a full load over the open countryside. It is extremely unlikely that any such arguments from me would have persuaded him to proceed, but Cherwell had promised that there would be no delays, and the Handley Page organization was soon proceeding on a very high priority to meet our requirements. We had asked for two bombers to be modified—one to be fitted with a hydraulic scanner and the other with a modification of the electrically driven scanner which I had developed with Metropolitan Vickers for the lock-follow AI. It now seems remarkable that on 27 March 1942—that is, only two and a half months after my visit to Handley Page—the first of these Halifax bombers V-9977, fitted with the experimental installation of a perspex cupola in the underturret position, landed at Hurn airport.

Less than three weeks after the Halifax arrived at Hurn, we were ready to test this first H_2S prototype in flight. In fact, on the evening of 16 April we had no success, because of a difficulty with the 80-volt alternator, which supplied the power, driven by one of the engines. The next morning's flight was successful, and from an altitude of 8,000 feet we could see the echoes from towns at ranges of four or five miles. Given that EMI had been chosen to engineer a production version of this equipment and that A. D. Blumlein was the leader of their team, it might be thought that our major problems had been surmounted. Unfortunately this was not so, since we confronted two major difficulties.

The first was an issue that had deep political repercussions. This was the use of the magnetron over enemy territory. The magnetron was comparatively indestructible, and it was considered that it would be a grave mistake to present the enemy with information on our advanced techniques in the centimeter field. An early order had in fact been given that the magnetron should not be used, but instead another type of microwave generator known as a klystron, the principles of which were believed to be known to the enemy, and which was in any case destructible. The klystron had been developed in the United States shortly before the war but the forms then existing could produce only a small amount of power. Urgent development was undertaken to make a klystron suitable for use as a transmitter, and in a few months klystrons capable of giving 5 to 10 kilowatt peak were produced. Many technical difficulties associated with the use of this tube were overcome, and by July 1942 prototype equipment existed that had interchangeable magnetron and klystron TR (transmit-receive) heads.

Before this stage was reached, there had been many bitter experiences with the magnetron H_2S equipment. The results in the Halifax were very much worse than the Blenheim experiments had led one to expect. This was mainly due to the higher altitude of operation and the consequent redesign and reposition of the scanner that had been necessary to give good coverage and all-around looking. The political pressure was so great that it was extremely difficult to settle down to satisfactory experimentation. On 6 May, Churchill had minuted the secretary of state for air, "I hope that a really large order for H_2S has been placed, and that nothing will be allowed to stand in the way of getting this apparatus punctually."[*] In spite of this political pressure the opponents of H_2S almost outnumbered its protagonists. The Americans began to claim that the successful British experience on town detection was not being confirmed in the United States; this, added to an unfavorable report from Group Captain Saye of Bomber Command, who wrote to the Air Ministry complaining that the display on the cathode-ray tube was like a "snowstorm," was mainly responsible for the meeting convened on 19 May 1942 by the assistant chief of air staff (ACAS) to review the whole question. At this meeting the air staff expressed faith in the development of H_2S and issued a series of directives—of which the two most important for us were (a) that the system should be accurate enough to guarantee that bombs would fall within an industrial or other area selected as a target; and (b) that the air staff would be satisfied in the first instance if the range of the device enabled

*Churchill, *Hinge of Fate*, p. 252.

the aircraft to home in on a built-up area from fifteen miles at 15,000 feet.

This much-needed vote of confidence in our development was quickly overshadowed by a series of events that changed my whole attitude to the tasks now confronting me. Suddenly the war began to press more heavily on our lives. In April, German ME-109's dropped bombs and machine-gunned our site at Worth Matravers (presumably unaware that we had moved from there the previous autumn). Then we were awakened early one morning by cannon fire, machine-gunning, and bombs, which destroyed the railway station and nearby buildings in Swanage. During the night of 24–25 April and the following night massive raids shattered the city of Bath. Suddenly a host of personal troubles made it even more difficult to concentrate on the problems of H_2S. Our parents and close relations lived either in or near the city. All communications were cut, and for some time our only information was that the city had been attacked by German bombers on two successive nights and that our defenses and night fighters had destroyed only four of the bombers. At last a message came from my own parents (they lived in the country outside of the city) that the family houses had been destroyed but that all the family were alive.

In those days we were allowed one rest day per week. By good fortune my allocated rest day at that time was on the day after the second attack on Bath and after the second night when no sleep was possible because of the roar of the bombers over our heads on the way to Bath. I drove to Bath to attempt what succor I could of the family. There are a few days of the war that I recall with vivid clarity nearly half a century later. This was one. The drive from the Dorset coast to the city of Bath is a mere seventy miles of unending beauty. On that lovely day the countryside through which I drove was covered with the spring flowers, and it was hard to realize that we were in the midst of a perilous war. That is until, within four miles of the city, I was stopped by a contingent of the army and told that no further progress was possible, because the city of Bath was closed.

Fortunately I had traveled the lanes and byways of that region of Somerset so often that I soon found my way by devious routes and, as I recorded in the diary, at last reached Bath at 4 P.M. and found it in a "most appalling mess." Eventually I negotiated the rubble to arrive at the home of Joyce's parents, where only a few weeks earlier our second child had been born in warmth and comfort. Her parents were there covered in grime, "in a window-less, doorless, and almost roofless house," the adjoining building having been flattened by a bomb. The surrounding area, in which I had spent so many happy hours with Joyce, "was just gone completely." A few hundred yards away was Beechen Cliff, on which had once stood the house of Joyce's grandfather. From there one could look over the city, and even if there were houses standing all seemed roofless. But the splendid Abbey in which I had so often been in the organ loft still rose magnificently over this scene of devastation.

Joyce's parents needed little persuasion to return with me to Swanage. As we drove out in the evening sunshine, great masses of people were thronging the roads, walking to sleep under hedges or anywhere where they felt safe from more bombs. "Poor Bath! destroyed as a working city; quite unbelievable. No trains, tele-phones, gas, or water." Over four hundred people had been killed and a thousand injured during the two nights. At last, I needed no further urging to do my utmost to help our own night bomb-ers. A strange psychology. I had seen the devastation in London and other cities, but this was acutely different in its impact. The romantic area of my young life had disappeared overnight—and for absolutely no military reason.

There was a savage irony about my brief visit to that shattered city—almost as though I needed that vision to surmount a great tragedy that was soon to engulf the H_2S development.

June 1942—The Crash of the Halifax Bomber

The meeting of the assistant chief of air staff at which faith in our H_2S development was expressed took place on 19 May. A week later we were forced to make an emergency move from our Dor-set laboratories and Hurn airport. The sequence of events leading to this emergency seemed to stem from the commando raid on the German radar station at Bruneval on the night of 27 February. This installation was believed to contain new radar equipment, and one of our colleagues, Donald Priest, with whom I frequently played tennis, had recently disappeared "on a secret mission," and eventually it transpired that he had been included in the raiding party to give guidance about the essential items of the radar station to be taken back to England. By mid-April there were rumors that a German parachute division had been assembled on the Cherbourg peninsula with the intention of making a retaliatory raid on TRE. Among the calamitous events that soon followed I never established whether these rumors were based on a real knowledge of German intentions. In any event, our situation became the subject of discussion in the Cabinet, and we were ordered to evacuate our site on the Dorset coast before the next full moon. We were situated on the Purbeck peninsula, approached by only one road through the historic settlement of Corfe Castle, a few miles inland. Rowe and his senior colleagues began a frantic search for new quarters. We were surrounded by barbed wire, guarded by the army and home guard, and every night our most secret files and all the magnetrons and associated equipment were driven inland behind this neck of the peninsula at Corfe Castle.

Early in May, and after many rumors about our probable destination, it was announced that we were to move to Great Malvern, in Worcestershire, take over the Malvern College buildings as our laboratories, and use the airfield at Defford, about seven miles east, for our airborne work. I had been sent to Malvern with a few senior members of TRE to investigate the facilities and the airfield, and on 26 May the whole establishment transferred from Dorset to Malvern. The facilities for our work did not exist; we were billeted on a largely hostile and puzzled local population who for security reasons could be given no enlightenment about the work of the establishment. The long queues for food at the Winter Gardens must still live in the memory of many who were involved in that extraordinary move.

Rowe had instructed me in January to form a new group for the H_2S development, and Dee had been elevated to the status of division leader with overall responsibility for my group and other groups concerned with the development of the centimetric AI and other systems. His division was allocated one or two floors of the science laboratory (the Preston laboratory) at Malvern College. There was one telephone in the whole of that science block, and someone posted a notice against it: "Beware! This phone is in parallel with Dee and Lovell."

These local and temporary difficulties were soon submerged in a great tragedy. The Halifax V-9977 had been flown from Hurn to Defford, where we were soon continuing our efforts to get a better picture from the H$_2$S at the operational altitude (15,000 feet) of the Halifax. On the weekend of 6–7 June we were meeting the EMI team at Defford to discuss its progress and development models. In fact, the second of the Halifax R-9490 aircraft had arrived at Hurn on 12 April and had been fitted with a Metropolitan Vickers electric scanner and the EMI prototype apparatus using a klystron. When this became airborne on 2 June, the same troubles with the polar diagram occurred. Naturally the EMI team was anxious to see how our own magnetron equipment with the Nash & Thompson hydraulic scanner was performing in Halifax V-9977. After a test flight with this equipment on the Saturday evening, I thought we were at last making some progress and arranged for two members of my group (Hensby and Vincent) to give a demonstration to the EMI team on Sunday afternoon, 7 June.

I do not need my diary note written in my distressed condition on the following day to remind me of the searing experiences of that Sunday. In this extract the reference to Ryle is to Martin Ryle, who had been placed in charge of a group to develop test equipment for the new centimeter equipment and was at Defford to discuss our requirements for H$_2$S.

Monday June 8 (1942): Yesterday was terrible. Our Halifax V9977 crashed and killed Hensby, Vincent, Blumlein, Browne, Blythen, S/L Sansom,* and a crew of 5 including the pilot Berrington. All day Saturday we were having a meeting [at Defford] with EMI. I arrived at

*G. S. Hensby was a civilian member of my H$_2$S group, Pilot Officer C. E. Vincent was an RAF member of the group, and Squadron Leader R. J. Sansom was the Bomber Command liaison officer on loan to the group. A. D. Blumlein, F. Blythen, and C. O. Browne were from EMI.

2.30 to continue with Ryle etc: they were just about to go off. They said they'd make it short. By 5 I began to worry. It took until 7.15 to stir up any interest in it. At 7.55 we heard it had crashed, at 8.35, 11 bodies were taken out, at 9, Group Captain King [the commanding officer at Defford] was driving O'Kane and myself down to salvage the apparatus. It was about 6 miles SW of Ross right in the Wye valley. It was a mass of charred wreckage, quite unbelievable. We salvaged some bits but there wasn't much except the magnetron recognizable. Arrived back 12.30 at Defford and finally Malvern at 2.

Many years later the loss of Blumlein in that crash was to receive some publicity. He was one of the most brilliant electronic engineers in the country and had been responsible for many of the fundamental ideas and patents that now form the basis of television. In my diary note at that time I wrote, "The loss of Blumlein is a national disaster. God knows how much this will put back H_2S." But he was not the only loss on that terrible June day. We had forged a marvelous working relationship with him and with Browne and Blythen. For me Hensby, too, was a tragic loss. At Birkbeck College, London, with H. J. J. Braddick, he had started research on cosmic rays before he joined the group. He was a fellow spirit, and on that depressing day of our arrival in Malvern I wrote, "Hensby and I found a pub, drank a pint and then at 10 p.m. climbed up the Worcester Beacon and decided it was a marvellous place to experiment on." The uninterrupted view eastward from the top of the beacon was later to lead to an idea destined to save the lives of many bomber crews, but not that of Hensby.

We were immersed in a desperate war, and the crash of a bomber on a secret flight received no publicity and only a routine investigation. Two weeks later I wrote, "Last week we tried to forget about the crash and get on with the work. This week we have perhaps succeeded better." On the fortieth anniversary of the tragedy efforts were made to mark Blumlein's achievements in the television field, and a memorial plaque was placed on the house where he lived.

But why did that bomber crash? My interest in that question was reawakened more than forty years later. The record in the Air Historical Branch of the RAF gives only a brief account of the crash, to the effect that the Halifax caught fire at a height of 2,000 to 3,000 feet because of the failure of the starboard outer engine, that the fire occurred when an attempt was made to restart this engine (the alternator supplying power to the H_2S equipment was on this engine), and that the fire extinguishers failed to work, probably because the bottles were not filled.

That might well have represented all our knowledge of the sequence of events had it not been for a remarkable investigation made over forty years after the crash by W. H. Sleigh, an aeronautical engineer who decided to make a detailed investigation after he retired from the staff of the RSRE (Royal Signals and Radar Establishment, formerly TRE) in 1984. Fortunately he located a number of key witnesses: Mr. Onslaw Kirby, a farmer who as a young man had witnessed the final moments of the Halifax as it crashed onto his farmland; Mr. Alex Harvey-Bailey, a former senior engineer of Rolls-Royce, the manufacturer of the engines fitted on the Halifax; and other Rolls-Royce engineers who were involved in the manufacture and test-flying of the engines during the war.

Sleigh's report, which he kindly sent to me on 2 September 1985, analyzed in detail the tragic sequence of events on that June day of 1942. It makes evident that the historical record is erroneous in that it would have been impossible for the pilot to have attempted a restart of the outer starboard engine and that the catastrophic sequence of events made the question of fire extinguishers irrelevant. Sleigh requested that the existing historical record should be amended with the following correct information:

Fatigue failure of an inlet valve stem caused by human error to ensure positive locking of that valve's tappet during a Squadron servicing check.

Loss of valve allowed combustion gases to ignite the charge mixture in the engine's induction and carburation system causing a major fire which in turn crossed the power plant's fire bulkhead and ignited the fuel in the mainplane fuel tanks.

The ensuing major fire caused the severence of the starboard mainplane at approximately 350 feet above ground level. The aircraft rolled to the right crashing inverted and exploded in a ball of fire killing all on board.*

In such a way the "unwitting oversight of an Air Force engine fitter to ensure the positive locking of one of the many tappet adjustment stud lock nuts during a routine safety check" caused these tragic deaths and led to a major upheaval in our affairs.

The Meeting with Prime Minister Churchill

Any feelings that this loss of the scientists and engineers and of our only airborne magnetron H_2S equipment would lead to the demise of the project itself were soon to be disabused. In the move to Malvern I had left Joyce in Swanage with the two young children, and three days after the crash I wrote to her, "The accident has left a great aching sore in our minds which is always there. We try to forget it by working hard and thinking that we just have to make a tremendous job of it to make up for that enormous sacrifice."

This mood was stiffened by the intervention of the prime minister. Ironically, on the day of the crash Churchill had minuted the secretary of state for air, "I have learnt with pleasure that the preliminary trials of H_2S have been extremely satisfactory. . . . We must insist on getting, at any rate, a sufficient number to light up the target by the autumn. . . ." In his earlier minute to the secretary of state for air on 6 May, Churchill had asked for "one man to be responsible for taking the necessary action by the proper dates and sending a monthly report." He had added, "I have heard Sir Robert Renwick mentioned as a man of drive and business experience. . . ."† On June 28 Sir Robert Renwick and his

---

*Annex 4 to Aircraft for airborne radar development by W. H. Sleigh. Unclassified RSRE (Malverne, U.K.) document. In his letter to the historical branch of the RAF, Sleigh pointed out that a failure of a combustion system in an aircraft engine is an extremely rare occurrence. When Sleigh was completing his report, an analogous sequence of events caused the tragic accident to a Boeing 737 at Manchester airport on 22 August 1985.
†Churchill, *Hinge of Fate*, p. 253.

aide, Frank Sayers, appeared at TRE. Renwick immediately became a person of immense influence in our affairs. He soon held the joint posts of controller of communications in the Air Ministry and controller of communications equipment in the Ministry of Aircraft Production. From 24 July he convened in the Air Ministry, every week, the series of "Renwick meetings" attended by the forty or fifty people who had to make the H_2S device work and to manufacture the equipment. Every day he would telephone me or Dee: "Any news, any problems?" The problems were dealt with either by Sayers or by means of Renwick's immediate access to Churchill.

Churchill believed that H_2S was vital to the bomber effort, and he had asked the secretary of state for air to take personal responsibility for the provision of this equipment. When the disaster of 7 June was reported to him, he summoned us to the Cabinet Room. On 15 June he had minuted the secretary of state for air, "I hope to have an H_2S meeting on Wednesday [17 June] at 11 a.m."* However, on 16 June, Churchill informed the king that he was about to cross the Atlantic to discuss urgent strategic issues, and not until the evening of that day did we learn that the H_2S meeting would be postponed. This was Churchill's second visit to President Roosevelt. Now that it is possible to cross the Atlantic in a few hours, it is salutary to be reminded that Churchill left Stranraer in a Boeing flying-boat just before midnight on 17 June and landed on the Potomac after being airborne for twenty-seven hours.

At this meeting with the president agreement was reached to pursue the work on the atomic bomb in the USA, pooling all existing UK resources and knowledge on the subject. Of more immediate impact was the telegram handed to Roosevelt during their meeting on the twenty-first. He passed it to Churchill: "Tobruk has surrendered, with twenty-four thousand men taken prisoner" (the number surrendering was later found to be thirty-three thousand). Churchill refused to believe this until he had direct confirmation from London. At this meeting, too, Churchill asked the president for three hundred Sherman tanks, which were only just coming off the production line, to be shipped to Egypt.

*Churchill, *Hinge of Fate*, p. 254.

78

However, it was to be October before this generous gesture helped in the defeat of Rommel at Alamein. When we eventually faced Churchill across the table in the Cabinet Room on 3 July, Dee and I found it hard to believe that he would spend his time on the H_2S problem, since only the previous day in the House of Commons in the debate, when we seemed on the point of losing Egypt, he had confessed to "the greatest recession of our hopes since Dunkirk."

Since the crash of the Halifax bomber on 7 June the pressure to get on with the H_2S had intensified, but we could do no flying. Our only magnetron H_2S had been destroyed, and the EMI team that alone knew the details of the equipment in the other Halifax had perished. It was against this background that, on the morning of 3 July, Churchill had assembled in the Cabinet Room the minister of aircraft production, Colonel J. J. Llewellin; the secretary of state for air, Sir Archibald Sinclair; the commander in chief of Bomber Command, Harris; the assistant chief of air staff, Cherwell; Watson Watt; Shoenberg, from EMI; and several others involved in the proposed production of the equipment and the deployment of the bomber force. Churchill was in no mood to listen to arguments. He immediately demanded that we should produce two hundred sets and have two squadrons in operation by October. He said they were "our only means of inflicting damage on the enemy." When reminded that we had not a single successful bomber equipped and that our only prototype, which was not performing satisfactorily, had crashed, and when Shoenberg said his firm could not possibly manufacture the equipment on that time scale, the prime minister turned to Cherwell, who to the consternation of commander in chief of Bomber Command remarked that the equipment could be built in the time on "bread boards." Supported by this assertion, Churchill dispatched us to an anteroom with the demand that we should not emerge until we had agreed how to produce the two squadrons by October.

The demand was impossible to meet, but it had the effect, in the words of my diary, of "starting the absolutely crazy pressure." For the remainder of that Friday and Saturday, Dee and I oscillated between Nash & Thompson (which had to make the scanners), EMI (which had to produce the electronics), and the

Office of the Controller of Telecommunications Equipment in the Air Ministry. On Sunday when I returned to TRE, my diary reads, "about a million people were at TRE, ½ a million being on H₂S."

Ironically the crash of the Halifax and the demands of Churchill had critical effects on the H₂S development. A week after the Churchill meeting, Cherwell came to Defford, and there on 10 July 1942 we persuaded him to let us concentrate on the magnetron and abandon the klystron version. Five days later the secretary of state ruled that the development work on the klystron for H₂S should cease and that the two H₂S squadrons should be equipped with magnetrons, but that the decision on their use should depend on the war situation—"If the Russians hold the line of the Volga." In fact, heartened by Cherwell's backing, we had already removed the klystron from the other Halifax R-9490, and on 12 July we made our first flight with a magnetron and the EMI prototype apparatus in that aircraft.

But none of this high-level pressure could make H₂S into a satisfactory operational device. The picture on the cathode-ray tube when flying the Halifax at 15,000 feet was poor. There were problems with the polar diagram that resulted in gaps in the ground returns, and only those of us skilled in looking at the tube could readily decipher towns from the general mass of ground clutter. Then on 5 July another face appeared at Defford. D. C. T. ("Don") Bennett, just promoted group captain, after a remarkable escape from enemy territory. On 27 April 1942 Bennett, then a wing commander, had led Halifax bombers in the earliest attack on the German battleship *Tirpitz*, which was moored under a precipice in an inlet of Trondheim Fjord, Norway. His aircraft was hit by anti-aircraft fire, but he and his second pilot escaped by parachute and landed in enemy-occupied Norway. After an arduous journey across snow-covered mountains and across icy streams, they reached Swedish territory and eventually made their way back to Britain. The Australian Bennett was not only very brave but also inexhaustible and dynamic. In 1938 he had set a new record for an east-to-west crossing of the Atlantic, and after the outbreak of war he led the first ferry flight of American and Canadian aircraft to the United Kingdom. For the next three years he became almost a part of my daily life. He maintained that the weather was best for flying at either 6 A.M. or midnight and demanded from the officer commanding at Defford that any servicing of the aircraft be done to suit his program. We no longer lacked flying hours in which to improve the system.

My own, small group of civilian scientists had, throughout these weeks, steadily been strengthened by service personnel. A matter of great importance was the appointment of Wing Commander (later Group Captain) Dudley Saward late in 1941 to take charge at HQ Bomber Command of the radio and radar aids for the command. The close relationship he established with us became of cardinal importance for the success of H_2S, and our flight testing was immensely eased by Flight Lieutenants David Ramsay and Peter Hillman, who struggled nobly under hazardous flying conditions to adjust baffles and other devices in the scanner, during flight, to minimize the gaps and the fades in the ground returns.

We could not meet Churchill's demand for two squadrons by October, but by 24 July we had produced a program for the fitting of twenty-four Halifax and twenty-four Stirling aircraft by the end of the year. Bennett was to use these for marking the target by flares in the newly created "Pathfinder Force," but during these autumn months we frequently wondered whether we had produced a device that anyone but a trained scientist could interpret. In my diary I wrote that it "amazed us that so much of the country's effort was being taken up to make these equipments when the only ones flying were terribly bad and almost unusable."

American Opposition

As we were attempting to improve the performance of H_2S and at the same time organizing the first fittings in the Halifax and Stirling bombers, my anxiety as to whether it could ever be used operationally was aggravated by American contacts. One of our former members, D. M. Robinson, had gone to America to help with the American radar developments. At the height of my own worries about the poor quality of the returns from built-up areas on the cathode-ray tube, Dee received a letter from Robinson in which he stated that "H₂S did not work in America." He had flown with a centimeter system and, although he had carefully manipulated the controls, had failed to detect any distinction between towns and ground. This cryptic announcement was soon followed by a visit from I. I. Rabi, the distinguished American physicist. In the autumn of 1940, after the visit of the Tizard mission carrying a sample of the new magnetron to the USA, the decision was made to form a group of American scientists "to translate the British 10 cm magnetron into a working radar system." Rabi became associate director of this powerful U.S. organization housed in the Massachusetts Institute of Technology and known as the Radiation Laboratory. Lee DuBridge was its director.

Because of intense security and wartime communication problems, the Radiation Laboratory scientists did not immediately grasp the significance of the TRE developments of centimeter radar. Thus, whereas we had detected an airplane with our first experimental centimeter radar in August 1940, the American group was not satisfied that this could be accomplished until it succeeded in February 1941—by which time we were only a month away from testing an airborne centimeter AI. In the case of the attempts to develop the centimeter airborne radar for use in bombers, the Americans had concluded that it was difficult to identify specific targets. In January 1943, when our own H₂S was about to be used operationally, the American group decided to abandon the 10-centimeter system and concentrate on the development of a shorter-wavelength, higher-definition system working on a wavelength of 3 centimeters.

In the meantime on 5 and 6 July 1942 Rabi and E. M. Purcell had visited TRE and, in a somewhat unpleasant meeting in Dee's office, had announced that in America they had concluded that our H₂S device was unscientific and unworkable and that if we persisted with our plans the only result would be that the Germans would obtain the secret of the magnetron. This depressing visit added to our own immediate problems because the American view was seized upon by the British opponents of H₂S, of which there were a number, both among the scientists and the operational RAF personnel. Even the day after the first successful operations, a letter from Robinson in America arrived in Dee's office suggesting that "the time was now ripe for settling the controversy as to whether this device will or will not work by doing some scientific measurements." The remarkable twist to this story, as I will later relate, is the subsequent turnabout of the American view when our own H_2S began successful operations.

The First Operational Tests with H_2S

During the autumn of 1942 the savage battles were being waged on the eastern front that would determine whether permission would be given to use the magnetron-equipped H_2S over enemy territory. We remained unhappy about the performance of the installation in the Halifax bomber, but nothing would stop the intense pressure to meet Churchill's demands. Renwick wielded his powers to such effect that before the end of August another Halifax and a Stirling bomber landed at Defford, complete with the underbelly perspex cupola and the internal racks to hold the equipment. It had been agreed that one of these test installations should go to the Bomber Development Unit (BDU) for use by RAF pilots and navigators, who would recommend whether or not the device was capable of operational use. On 30 September we sent the Halifax to this unit with the test installation of the production units that would be fitted in the first operational squadrons. I have a photograph of the cathode-ray tube display taken in this Halifax when flying at 10,000 feet toward the towns of Gloucester and Cheltenham. There is a vacant gap of several miles immediately underneath, then a disconcerting circular smudge of ground returns with the radar responses from the towns about ten miles ahead appearing as indistinct areas on the tube with another response midway, which we discovered to be an airfield. Once again I wondered what use the conventional navigator could make of such a picture and would not have been surprised if the BDU had soon returned the Halifax as useless. It did not do so, and at the end of November its report to the commander in chief of Bomber Command was so favorable that Air Marshal Sir Robert Saundby ("Bomber" Harris's deputy) asked the Air Ministry for permission to operate over enemy territory with this equipment as soon as Bennett's Pathfinder Force had one squadron equipped and trained.

In the meantime I had a vivid experience that made me realize how urgently the RAF needed this equipment, even in the current imperfect state. During the second week of October, Saward had landed a small plane at Defford to fly me to the BDU airfield at Gransden Lodge to talk to the navigators who had been using the test equipment in the Halifax. They seemed content with the performance, and that night I understood the reason. We flew on to an operational airfield at Lakenheath. In the evening the Stirlings of that squadron departed for a raid on Cologne. I was at the end of the runway with Saward, and as the roar of the bombers faded I asked Saward how they could possibly find Cologne in a darkness that made it hard for us to find our way back to the mess. He was not hopeful—Gee, the ground-based navigational aid that just covered the Ruhr from the east coast, had been jammed. His pessimism was soon justified. In the early hours of the morning the crews that had survived the onslaught by the German defenses began to filter back to the interrogation room. One man after another confessed that he had never even seen the target and that "all was hellishly confusing over northwest Germany."

Like the sight of Bath after the bombing of April, that night restored my flagging spirits. The gaps and the fades in the responses on our cathode-ray tube were of little consequence to the navigator who had no means but the stars and dead reckoning to find his way in a bomber being raked with ack-ack fire and constantly weaving to escape the night fighters. I wrote in the diary that I was now "absolutely aware of the vital necessity of H_2S. Went to bed at 3 am and got up again at 7 am to fly back to Defford."

Now there was no time to worry about improving the presentation. The activity on which we were engaged became known as a "crash program." We had been a research group; now we were a fitting and test group. The bombers flew in to Defford; the fifty complete sets of equipment came from EMI as had been promised; we fitted them, made a flight to test the system, and dispatched the bomber to Bennett's Pathfinder Force. By the end of the year (two months behind Churchill's demand) there were twelve H₂S-equipped Halifaxes and twelve Stirlings in the Pathfinder Force.

The Bombing of Hamburg

On 8 December the secretary of state for air and the Combined Chiefs of Staff agreed that these H_2S aircraft could be used operationally over Germany at any date after 1 January 1943. The criterion set in July for the use of the magnetron—that the Russians should hold the line of the Volga—had been met in one of the decisive battles of history. In mid-September the German armies had reached the outskirts of Stalingrad, but, as Churchill described the battle, "the battering-ram attacks of the next months made some progress at the cost of terrible slaughter. Nothing could overcome the Russians, fighting with paranoiac devotion around the ruins of their city." A month later "General Paulus's Sixth Army had expended its efforts at Stalingrad and now lay exhausted with its flanks thinly protected."[*] At the time of the decision about the operational use of the magnetron, on 8 December, the Russians had delivered their encircling assault and Paulus's army was trapped between the Don and the Volga.

Although Bennett had the H_2S-equipped aircraft, the crews still had to be trained, and this task was encumbered by the poor serviceability of our equipment. O'Kane had gone to Bennett's headquarters at Wyton, near Huntingdon, to assist with the preparations, and on the evening of 29 January he phoned and remarked that it might be a good idea for me to take the weekend "in the country." On Saturday morning a member of Bennett's staff phoned and told me to "come immediately." I was in the mess at Wyton by five-thirty that evening. H_2S was to lead a raid on Berlin, but because of weather conditions the target was changed to Hamburg. Shortly before midnight seven H_2S Stirlings and six H_2S Halifaxes took off to mark Hamburg for the main

*Churchill, *Hinge of Fate*, pp. 524–25.

85

bomber force. There was no sleep on that night of 30–31 January 1943. The weather conditions were so bad that forty Lancasters of the main bomber force had turned back, as had four of the H$_2$S aircraft, with their sets out of action. At last as dawn was breaking, an excited navigator of a Halifax landed. He had been able to identify the Hamburg docks on his H$_2$S screen, and he was soon followed by another Halifax and four Stirlings, which had all reached the target with their H$_2$S working well. Saward was at another of the Pathfinder airfields, where the returning crews were excited at the H$_2$S picture of coastlines, estuaries, and rivers appearing on the tube like a well-defined picture of a map. The navigator who reached the Hamburg docks reported that they appeared as "fingers of bright light sticking out into the darkness of the Elbe."

A few nights later, 2–3 February, ten H$_2$S planes of the Pathfinder Force marked Cologne; the next night eleven H$_2$S aircraft marked Hamburg again and the night after that Turin. Our anxieties were at least temporarily assuaged. After these four raids HQ Bomber Command reported, "H$_2$S is the most successful blind navigation and bombing aid yet devised." There was a signal from Bennett of the Pathfinder Force on 4 February: "Heartiest congratulations from myself and the users to you and your collaborators in the development of the outstanding contribution to the war effort which has just been brought to action."

If I had not retained a diary of those days, I would have remembered them as ones of elation at the success of our efforts. But it was not so. Inevitably one of the H$_2$S bombers was soon lost over enemy territory, and we knew that the Germans could not fail to acquire the secret of the magnetron and of H$_2$S. I could not know then that thirty-four years later I was to learn the fate of that equipment. On 19 April 1977, when I was on a liaison visit to our German colleagues using the large radio telescope at Effelsberg, near Bonn, I dined with the director, Professor O. Hachenberg. He asked me what I had done during the war. I told him. And did he, too, remember those years? "I was also a young man working on radar, and one day in February 1943 I was sent to investigate the equipment in a bomber which had been shot down near Rotterdam. It was your H$_2$S." Was he surprised to discover the secret of the magnetron? "Not at all; we had already made a similar device, but we could not discover how your system worked. There was a broken glass tube which completely mystified us." The tube was filled with an inert gas at low pressure and arranged so that a spark gap, operated by the transmitted pulse, short-circuited the input of the receiver, thereby protecting it from damage. This was a simple form of TR (transmit-receive) device that enabled the same antenna to be used for transmitting and receiving.

We had been driven relentlessly by the political and operational pressure. We had been optimistic that the device would in some way be useful. In the days ahead we were to need all our optimism. After the immediate release from the anxieties and problems of these first H_2S operations, my diary entry for 13 February reads, "Strange but I can't feel a bit elated, only weighed down with the vastness of things in hand and the future." After the night of those first operations I drove back to Malvern. It was a Sunday. We always had to work on Sundays, and it was a very long time since I had been inside a church. The device for which my group had been responsible had already guided our bombers to kill many people. As I write about this forty-five years later, I wish I could say that I had gone into the ancient priory church at Malvern on that Sunday evening. I did not do so. "I came back in the [Sunday] evening feeling devilishly tired . . . a grand feeling to have done something which had gone over, but . . ." The *but* in that sentence leads me to another burden that had descended on me and my group, and I must now write about this—a saver, not a destroyer, of lives.

Centimeter ASV against the U-boats

". . . but trouble at Chivenor with Coastal." The airfields of the Pathfinder Force were in the east of England. Chivenor was an airfield in North Devonshire, many hundreds of miles to the west. Throughout January, as the Pathfinders were preparing for their operations with H_2S, my own troubles were divided between them and the Coastal Command squadron at Chivenor.

By the summer of 1942 the sinkings of merchant shipping in the North Atlantic by U-boat attacks had become a very grave issue. In *The Second World War* Churchill wrote that the losses of our merchant fleet, reaching over 600,000 tons *per month*, by mid-summer "constituted a terrible event in a very bad time" and that the "U-boat attack was our worst evil."* Indeed, any history of that period reveals the critical nature of the U-boat attack on the ships bringing oil and essential supplies to Britain.

When I became involved in these affairs late in 1942, I was told that Churchill and the British Cabinet had the very gravest anxieties about the onslaught to be expected during the summer months of 1943. Churchill had repeatedly begged President Roosevelt for help, but this aid from the United States in the form of destroyers and long-range aircraft operating over the North Atlantic had so far barely compensated for the ever increasing size of the U-boat fleet able to operate on the Atlantic routes, and Churchill's great fear was that Hitler would decide to stake all upon the U-boat warfare. In my later years I was to derive moral compensation for the destruction wrought by H_2S by the part it played in this battle in 1943. In Churchill's words, "H_2S was adapted for use in the ASV role with striking advantage. In 1943 it made a definite contribution to the final defeat of the U-boats."† I must now describe how this came about, so that when the Pathfinders were using H_2S to bomb Hamburg at the end of January 1943 I was in a tug-of-war between Bomber and Coastal commands, each of which was fighting for our few pieces of H_2S equipment and for our help.

In 1940 the Germans had captured the French west coast ports and established submarine bases on the edge of the Atlantic. This presented a new and serious threat to the Allied convoy routes. The Bay of Biscay became a "transit area," that is, an area through which the U-boats passed to and from their bases but which was not primarily their operational area. Fortunately this transit area was within reasonable range of aircraft based in Britain, and a violent and lengthy battle between aircraft and U-boats soon developed.

*Churchill, *Hinge of Fate*, p. 110.
†Churchill, *Hinge of Fate*, p. 254.

ASTRONOMER BY CHANCE

Until the summer of 1942 the U-boats working from the bay ports had matters largely their own way. If they felt in danger of attack from aircraft in the daytime, they submerged, surfacing at night to recharge their batteries. The ASV system that existed at that time in the night-flying aircraft was essentially the one that Taffy Bowen was developing when I joined him at Scone airport in 1939 and that we fitted into the Hudson aircraft at St. Athan later that year. It worked on a wavelength of 1.5 meters and could detect submarines at night at ranges of several miles and enabled aircraft to home in on the target. It was necessary, however, to carry out the final depth-charge attack visually (among other reasons, to be sure that the target was actually a U-boat), and, except under occasional full-moon conditions, a satisfactory final visual sighting could not be obtained. Hence the U-boats were not seriously hampered in their passage to and from the bay ports, and a considerable fleet was maintained in the Atlantic, menacing the convoy routes and causing the merchant shipping losses to reach a very high figure—more than three million tons of shipping were sunk in the Atlantic Ocean alone between January and July 1942.

In June 1942 an important development was brought into operational use. A powerful searchlight with a flat-topped beam and with an azimuth spread of about 11 degrees—the "Leigh light" (named after the originator, Wing Commander H. Leigh)— was mounted in a retractable cupola underneath the fuselage of Wellington aircraft fitted with the ASV apparatus. Then, in the final approach to the target, located and homed in on by the ASV apparatus, an operator in the nose of the aircraft could switch on the searchlight when about one mile from the target. If the target was identified as a U-boat, the attack could be carried out under conditions of full visibility. This sudden illumination and attack during what had previously been regarded as "safe" hours upset the U-boat crews so much that although the total number of night sightings during June and July was only about twenty, by August the U-boats were no longer surfacing at night for recharging purposes, but during the daytime. This enabled the day forces of antisubmarine aircraft to be deployed in daylight. Statistically this was equivalent to nearly 100 percent sightings of U-boats known to be operating in the area of the bay.

The German scientists quickly provided the U-boat crews with a successful countermeasure in the form of a simple radio receiver known as Metox. This receiver enabled the U-boat crews to pick up the radiations from an ASV-equipped aircraft long before it was at the range when it normally operated its Leigh light. Thus the U-boats could submerge when their receivers indicated that an ASV transmission was homing in on them. The density of the aircraft patrols was insufficient to make the U-boats submerge so frequently that it became a nuisance to them, and the pre–June 1942 conditions were rapidly restored. As a result, during the winter of 1942 and early months of 1943 the sightings of U-boats in the bay by night and day were very few, and after a "seasonal" decrease shipping losses took a sharp turn upward in January and February 1943 and showed signs of reaching unprecedented proportions.

The problem of the U-boats' listening to the aircraft ASV transmissions had one obvious technical solution, at least as a temporary measure—namely, to make a major change in the wavelength of the ASV transmission. Such a system, working on a wavelength of 10 centimeters, had in fact been under development by another group in TRE. This group did not have the priority attached to the H_2S development, and by September 1942 it was obvious that any ASV 10-centimeter system emerging from that group and the manufacturers (Ferranti) would be far behind the equivalent apparatus to be produced for H_2S. Of course, the competition between the various groups in TRE was intense, but on 25 September 1942 Renwick once more summoned us to London and wielded his authority to rationalize this situation. The existing ASV group was disbanded and the arrangements with Ferranti canceled. In that manner part of the previous ASV group became my responsibility. This is one example of the competitive spirit within the wartime establishment that undoubtedly stimulated individuals to ever greater efforts. Occasionally the disputes erupted in a practical fashion—notably on the epic occasion when Martin Ryle became so irritated in a meeting with his divisional leader (Robert Cockburn) that he seized an inkpot from Cockburn's desk and hurled it at his head. The mass of ink on the wall behind Cockburn's head remained for many a year.

The disruption of the ASV group naturally caused a great disturbance, and the commander in chief of Coastal Command (Sir Philip Joubert de la Ferté) and his chief signals officer (Group Captain P. Chamberlain) soon came to TRE and demanded a demonstration of their "new" equipment. On 27 October we gave them a very poor demonstration using a Stirling bomber fitted with H$_2$S, and their antagonism increased. Presently there were to be radical changes in the Coastal Command personnel, and when Air Marshal Sir John Slessor succeeded Joubert as commander in chief of Coastal Command my relations with him and his staff became cordial. Even so, in spite of the unhelpful attitude and antagonism of Joubert and Chamberlain, by December 1942 we had two Coastal Command Leigh light Wellington aircraft fitted with modified H$_2$S equipment. The scanner had to be changed radically. In the bombers it was under the fuselage. In the Coastal Command Wellingtons that position was occupied by the Leigh light, and a new perspex nose thus had to be fitted to the Wellingtons so that our scanner occupied that forward position. There were endless problems in integrating this equipment with the various Coastal Command devices, but by the end of January 1943 we had three of the aircraft at Chivenor and a dozen a month later. Seven from my own group were diverted there to help maintain the equipment and train the operators. A decision had been made to include two of these centimeter ASV aircraft in the Bay of Biscay patrol by the Chivenor squadron on the night of 1 March. The station commander (whose name, fortunately, I did not record in my diary) was even more antagonistic than Chamberlain. I can never forget the wintry dawn of that March day as we were awaiting the return of the aircraft. Fearful that some accident might have befallen the ASV-equipped plane in which R. L. Fortescue, one of the TRE people, was operating the radar, I waited with the station commander in his office, as bare and bleak as the view of the airfield through the window. To him we were just a nuisance: "It's all a matter of joss, just joss. What good do you think your miserable apparatus is going to do over the vast space of the bay?" He was soon to learn and be dispatched to a place where his obstructionism was of less consequence. Under the strict military discipline of war such dismissals

were relatively simple to arrange—the individual was posted elsewhere, often to a nonoperational position.

Eventually the aircraft returned with nothing to report—except
that the ASV had been serviceable. Fortescue, who had been
made a temporary RAF officer wearing a uniform, offered to fly
again the next night and, although no U-boat was sighted,
endeared himself and the equipment to the crew by detecting an
approaching fighter and giving the pilot instructions for evasive
action. Otherwise for two weeks the operators of the ASV had little to report. During the night of 17 March, however, the equipment detected a U-boat at a range of nine miles—but the Leigh
light jammed and no attack was made. On the eighteenth a U-
boat was sighted again and attacked with six depth charges. The
crew reported that the "U-boat was fully surfaced and under way,
showing no signs of suspecting an attack." A disastrous period for
the U-boats had begun. The few aircraft equipped with centimeter ASV operating from Chivenor detected thirteen U-boats before
the end of the month and a further twenty-four in April. The U-
boats could not face these sudden and unsuspected attacks during
the hours of darkness. They were forced to adopt a change of tactics and remained submerged during the hours of darkness and
surfaced by daytime. Admiral Karl Dönitz ordered the U-boat
commanders to stay on the surface and fight it out with the daytime air patrols. It was a disastrous order, for by this time the
long-range daylight patrols operating from America had closed the
unpatrolled gap that had hitherto existed in mid-Atlantic because
of the limited range of the patrols from the United Kingdom. On
the average, in May, every U-boat crossing the Bay of Biscay suffered at least one attack. In March the U-boats sank 700,000 tons
of Allied shipping. Instead of rising to the fearful peak expected
in the summer of 1943, the tonnage lost had by July fallen to less
than 100,000 tons per month. A few pieces of the H_2S equipment
modified to serve in the ASV role had in a few weeks led to a
remarkable transformation in the strategic situation, causing Hitler to complain, "The temporary setback to our U-boat campaign
is due to one single technical invention of our enemies."[*]

*From a radio broadcast by Hitler in late summer of 1943, cited in A. P. Rowe,
One Story of Radar (London: Cambridge University Press, 1948), p. 160.

Indeed, we had expected the setback to be a temporary one, because there seemed no reason why the Germans could not quickly provide the U-boats with a listening device in the centimeter band as they had done for the 1.5-meter ASV. Late in 1943 we had evidence that the Germans were using such a receiver, known as Naxos. As I shall describe, circumstances with the H_2S had led us to the urgent development of a 3-centimeter version, and we had tried to arrange for a conversion of this equipment, to be made available to Coastal Command, so that once more the U-boats would be foiled in their attempts to detect an approaching aircraft. Toward the end of October, Air Marshal Sir John Slessor, now the commander in chief of Coastal Command, came to see me in TRE to make sure that everything that could be done to meet this new threat was being done. For our part we had developed and tested in flight a powerful 3-centimeter version of H_2S for fitting in the Coastal Command Wellington aircraft. The procurement and fitting arrangements for two hundred pieces of this equipment had been handed over to the appropriate people in the Air Ministry and the Ministry of Aircraft Production, but the weeks of delay had seemed like a sabotage of our efforts.

Slessor arrived on a Saturday afternoon and, on hearing my account of this affair, quietly removed his beribboned jacket, asked for paper, and proceeded to draft a five-page foolscap letter to the secretary of state for air, calling for disciplinary action against those responsible for the delay since, unless his command could have this equipment, the success of the bay offensive would be "hopelessly prejudiced." We had already been waiting two months for the special operational aircraft which we were to equip with this new 3-centimeter ASV. "Their repeated attempts to make progress in this direction have, however, been met with the astonishing claim that there is no stated operational request . . . and therefore that nothing can be done to get on with the trial installation. . . . This is either crass stupidity or pettifogging obstructionism of the worst order, which is likely to have most serious results on the conduct of the Anti-U-boat War. . . ." Slessor then gave the reference to his clearly stated operational request (of April 1943) and continued that "no-one but a congenital idiot" could imagine that our 3-cm equipment would be of any value if there was no appropriately fitted antisubmarine aircraft to use it.*

The effect of this letter delivered on a Sunday to the Air Ministry in London illustrates the very close working liaison we had in those critical days with the operational staff, and also the antagonism between Bomber and Coastal commands, the former believing that our new equipment would be better used to locate and bomb the U-boat pens in the bay ports. Three days later I was sandwiched between Slessor and his staff on one side of a long table in the War Cabinet offices, and the Bomber Command staff on the other side. The deputy chief of air staff was in the chair and, after hearing the arguments, ruled that Coastal Command should be given thirty pieces of the new equipment immediately. I drove back to Malvern feeling contented, to be met at the gates of TRE by an irate Rowe accusing me of failing in my mission to get the equipment to Coastal Command. "But the Air Staff have agreed to provide Slessor with thirty of the 3-centimeter-equipped Wellingtons."—"Ah, but since you left London, Harris [the commander in chief of Bomber Command], has phoned the Prime Minister and got the decision reversed."

So Slessor and Coastal Command did not get the 3-centimeter ASV operational by Christmas 1943, as we had promised. Months later it was operating over the bay, but fortunately the U-boats never recovered from the violence of the attacks to which they had been subjected during the summer months of 1943. By August 1944 the Allies had captured the bay ports, and the battle of the Bay of Biscay into which I had been unexpectedly drawn became a matter of history. There were to be many criticisms of H_2S and of the policy for its use. Those who were so anxious to criticize the folly of the great priority we had been allocated conveniently overlooked the fact of its conversion to help Coastal Command. No other centimeter equipment could in March 1943 have been operational over the bay to play that vital role in saving many hundreds of thousands of tons of merchant shipping from the U-boats.

*Letter from Sir John Slessor to the Under Secretary of State, Air Ministry, 24 October 1943. Collection of the author.

The Improved Versions of H₂S

The original 10-centimeter H_2S equipment used by the Pathfinder Force in the early months of 1943 succeeded because of the enthusiasm of Bennett and the skill of his picked navigators. The method of its use by a specially trained group—marking the target for the main bomber force—aroused much criticism and antagonism from others in the command who claimed they could navigate and bomb with greater precision. As the nights shortened, there was insufficient darkness for the bomber force to penetrate deep into enemy territory, and the bombing operations were concentrated on the Ruhr. This area was within range of the device known as Oboe, in which a few aircraft could be controlled from ground stations in Britain and make it possible to mark or bomb a target with far greater precision than was possible with H_2S. Oboe had been developed by a group in TRE working on the floor below my own group, and in the summer of 1943 a good deal of political and scientific opposition to H_2S emerged.

For our part we had worked continually to improve the H_2S display. There were two major problems. First, the existing display was very uneven. We had great difficulty in producing a polar diagram from the scanner under the fuselage that would give uniform coverage of the terrain over which the aircraft was flying. There were gaps and fades in the coverage, and as the Lancaster bombers came into service the situation was aggravated because of the greater operational height of these aircraft compared with that of the Halifax and the Stirling. Today, whenever I hear a young scientist complain about his working conditions, I wish I could transport him to the cramped cupola of a Lancaster flying at 20,000 feet and instruct him to experiment with the shape and feed arrangements of the rotating scanner. Fortunately, the two young RAF flying officers in my group, Peter Hillman and David Ramsay, willingly bore the brunt of this experimenting. By the spring of 1943 they had made a very great improvement in the H_2S picture, fitting a shaped waveguide feed to the scanner instead of the original dipoles—and they richly deserved the Air

Force Crosses subsequently awarded to them. Forty years later I answered the telephone—it was Hillman, who had returned from Australia to see his friend Ramsay again. They came to my office, and I reminded them of those days when they worked without oxygen at high altitudes to make H_2S a better system. "Oh, it was nothing, it was a pleasure." They were typical of the fifty-odd civilians and servicemen I had then—Americans, New Zealanders, Canadians, and Australians—who willingly responded to every call.

The new feed system devised by Ramsay and Hillman was produced and fitted into the Pathfinder Force aircraft by July. On the nights of 24, 27, and 29 July, Bennett's planes using this new equipment marked Hamburg, and eight hundred heavy bombers dropped their bombs on the flares to reduce the city to ruins in a great fire storm. The subsequent reconnaissance photographs taken by the high-flying Mosquito aircraft revealed the utter destruction of thousands of acres of the city. The Admiralty had placed great pressure on Bomber Command to attack Hamburg because of its extensive shipbuilding yards, responsible for nearly half of the German production of U-boats. The city was estimated to contain three thousand industrial establishments of vital importance to the German war effort. After the attack Albert Speer, the armaments minister, warned Hitler that a series of attacks of this ferocity and accuracy extended to six more major cities would bring Germany's armaments production to a halt. The success of H_2S in leading these attacks silenced its opponents—at least temporarily. It was twenty years after these attacks before I went to Hamburg. Then I was consoled to see the beauty of the re-created city and relieved that the H_2S, in spite of accusations to the contrary, had not spread the attack to a nearby hamlet and engulfed the church and the organ on which Bach had played.

The second major problem, even with the improved picture, was the relatively poor definition. From 20,000 feet the nucleus of the major cities would often appear as an unresolved area of brightness on the tube. The solution was obvious—reduce the wavelength to 3 centimeters, which, with the same-sized scanner under the aircraft, would improve the definition by three times. In January of 1943, even before the first operations with H_2S, we had begun to modify a 10-centimeter H_2S for operating on 3 centimeters, and on 11 March this system had its first flight test, in a Stirling bomber. Early in June I was astonished to learn that a powerful contingent from America, including Lee DuBridge, the director of the MIT Radiation Laboratory, had arrived in London in the hope of persuading the Air Ministry to fit their own newly developed 3-centimeter H_2S (H_2X in the USA) into the British bombers. That was less than a year after the unpleasant visit of Rabi and Purcell demanding that we stop the development of the "unscientific and unworkable" project.

In common with Dee and many others in TRE, I was greatly provoked by this move—but scarcely so much as Saward and his colleagues in Bomber Command, who were well aware of the relative lack of operational experience then existing in America and of the vital importance of the intimate liaison between us and those using the equipment on operations. Fortunately, in two meetings on 7 June with DuBridge and his party, chaired by Renwick and with Cherwell present, the decision was made to proceed with our 3-centimeter H_2S and to equip three squadrons by the end of 1943.

However, our steady progress toward that target was soon to be in turmoil because of an urgent operational problem. By late August the nights were long enough for Bomber Command to stage a major attack on Berlin. It did so on 23–24 August and 31 August–1 September, with poor results and near disastrous losses of aircraft. Over 1,300 heavy bombers were used in these two attacks, and 104 were lost. H_2S, which by then was in a part of the main bomber force as well as the Pathfinders, became the target for much criticism—the poor results were alleged to be due to the inaccuracy of the H_2S picture.

Saward was soon in my office with a photograph of the cathode-ray tube taken in an aircraft near Berlin, and I realized there was some justification for the criticism. The entire 10-mile "bombing scale" was covered with blurred responses from the city, and there was nothing on which the navigator could make a fix of the aircraft's position. Then Saward produced a map of Berlin, and it required little imagination to see that an H_2S with better resolution might reveal the lakes near the city and provide accurate fixes for the Pathfinder Force. The commander in chief of Bomber Command recognized this too, and his deputy Saundby had already been in TRE demanding some of the 3-centimeter H_2S by November. That was a hopeless prospect—the "official" program for equipping the Pathfinder Force by the end of the year seemed most unlikely to meet even that target.

Stimulated by Saward and Bennett, and with the connivance of Renwick, we had by mid-September made, in Saward's words, the "craziest plans possible" to fit six Lancaster bombers with this equipment by early November. Saward was to send us the six bombers, which we were to equip at Defford. Len Killip, an experienced operational navigator and then a flight lieutenant, was lent to my group, and he relentlessly flew with the experimental system in the Stirling bomber with members of my group, and steadily the performance was improved. The wall of my office was covered with a huge chart with jobs allocated every day for every member of the group. Indeed, it was an extraordinary plan made only with the secret agreement of those involved and with no kind of official backing. I think now, more than forty years later, that only in the heat of battle can one make such plans—and succeed. "November 11th—tremendous pressure on the six Lancs. I have a feeling that Bomber Command are about to make an all out effort on Berlin. Told Bennett last night that we would try to deliver him 3 Lancasters on Sunday and 3 on Tuesday."

We fulfilled our promise. Bennett had his six Lancasters by 17 November with a large slice of my group to maintain the equipment and instruct the operators. Before me as I write now I have the photographs of the H_2S picture taken from one of those 3-centimeter H_2S Lancasters during the run up to Berlin at 19,000 feet. The thin line of the Tegeler See and Wannsee stretching southwest from the city appears as on a map, and so does the Templehof airport. The crews claimed that their markers went down within one hundred yards of the assigned target. These sustained attacks on Berlin from mid-November 1943 to February 1944 had, in Saward's description, "staggering effects and brought howls of anguish from Goebbels, the Nazi Minister of Propaganda, and from Albert Speer, the Minister of Armament Production."[*]

*D. Saward, *Bernard Lovell* (London: Robert Hale, 1984), p. 100. See also D. Saward, *Bomber Harris* (London: Cassell, 1984), pp. 221–23.

When the records of these raids became available, it was evident that almost every major industrial works in the Berlin area had been destroyed, justifying Goebbels's implied tribute to the H_2S that "the English aimed so accurately that one might think spies had pointed their way." In January 1983, when my thoughts were far from Berlin and those events of thirty-nine years ago, I received a letter from a resident of the United States, who as a boy was in Berlin at that time. He wrote, "The realization that the English could see through the clouds caused nothing short of consternation in the Air Ministry, where my father heard the news."

Fishpond

The Worcester Beacon is the highest point on the ridge of the Malvern hills. In summer it shades Great Malvern from the afternoon sun, and this beautiful spot can be reached after an easy ascent from the town. Dudley Saward was a frequent visitor and often stayed the night with us at a flat we had secured in Great Malvern close to the college to which TRE had been evacuated in the spring of 1942. One night in mid-April 1943, nearly a year after this move from Swanage, when food rationing was making it difficult to provide food for our visitors, we drove around the beacon to dine at the Westminster Arms, on the west side of the hills and even closer to the ridge. Dudley wanted to walk back to the flat over the beacon, and we left Joyce to return in the car. I sensed that he was worrying about something other than the behavior of the H_2S sets that the Pathfinders were now using regularly and that were now being fitted in the main bomber force.

We reached the cairn at the top of the beacon as daylight was fading, and as we paused there I discovered why Saward had wanted this lonely walk homeward. He was anxious about the number of bombers being attacked and destroyed by German night fighters. They carried a tail warning device that had lost its effectiveness because of a change in the night-fighter tactics, and the centimeter AGLT (Air Gun Layer Turret) that had been developed in Dee's division by Hodgkin could not be in production for

99

another year. It was a depressing story, especially since the opponents of H_2S were arguing that the night fighters must be homing in on the H_2S transmissions. Saward told me that the German night fighters were now attacking our bombers from underneath and astern. As we looked eastward, I recalled my visit there nearly a year earlier with Hensby. He had remarked on the view and mentioned his belief that there was no high ground between us and the Urals. The transition to the idea that there would always be at least 15,000 feet of empty space before useful H_2S echoes appeared was immediate—and we were wasting that part of the H_2S picture as of no interest. I asked Saward if he would send me a mechanic to build a special display unit. We would expand the center region of our display and give the rear gunner a special cathode-ray tube on which I thought he would see an approaching fighter in the region before the ground returns of H_2S were manifest.

Within a few weeks we had tried out the idea in flight at Defford, and a mock attack by one of our fighters was easily revealed as a moving spot on the screen inside the ring of the H_2S ground returns. On 27 May I told Saward we were ready to give him a demonstration. A Mosquito fighter made mock attacks on the bomber, and Saward was so impressed that, at the urgent request of the commander in chief and his deputy, Renwick two days later summoned a meeting at which the manufacturers were ordered to produce thousands of this "Indicator Unit Type 182." The device was allocated the code name Fishpond, and in October, only six months after my lonely walk with Saward on the Malvern hills, it was used over Germany. A month later one-half of all the H_2S sorties used Fishpond, which was destined to save the lives of many hundreds of our bomber crews.

Scent Spray and the American Fortresses

By the spring of 1943 the production line of H_2S for the main bomber force had commenced, and we no longer depended on the small number that had been produced in the "crash program" for the Pathfinder Force. The prospect of the availability of large numbers of this H_2S system soon had other consequences. On 21 March we received two visitors from Combined Operations who asked about the possibility of modifying H_2S for navigating and ranging with tank landing craft. On 19 April I was asked to modify equipment for a demonstration. Three weeks later, on 9 May, we sailed from Portsmouth in the landing craft LCT-162 that we had fitted with the modified H_2S with the scanner in a perspex housing 36 feet above sea level. Commander R. T. Paul and Lieutenant H. F. Short of Combined Operations were on board; the low-lying coastlines appeared on the cathode-ray tube at ranges of ten to fifteen miles and the Nab Tower at five and a half miles. My diary note of that day reads, "There was nothing wrong with that Sunday, everything was quite perfect, and it took only a few days before the Chief of Combined Operations was sailing in LCT-162 and TRE were engaged in the modification of 20 sets." On 10 July these were used in the invasion of Sicily, and the job of the large fitting program passed to the Admiralty. Our "Scent Spray" soon became the Admiralty Type 970, and the photographs before me of the cathode-ray tube taken from a tank landing craft as it approached the French coast on D-day, June 1944, are impressive reminders of the consequences of the great priority accorded to our H_2S.

Another consequence of the large-scale production of H_2S in 1943 involved the American Eighth Bomber Command. Although equipped with excellent precision visual bombsights, the American air force soon discovered that the opportunities for bombing Germany from clear blue skies were few and far between. The commanding officer, General Ira Eaker, made urgent requests for his Flying Fortresses to be fitted with 10-centimeter H_2S. Arrangements were made at Defford on 18 March, and in June a Flying Fortress landed there for our trial installation. For us this was an unsought and difficult burden. The Fortresses' operation at 30,000 feet caused new problems with the coverage given by the scanners used at the much lower operational height of the Royal Air Force bombers. The constant high-altitude flying soon sent Hillman to the hospital with pneumonia, and a corporal in my group lost his life through lack of oxygen and extreme cold.

An awkward problem was solved by a vigorous Canadian officer attached to my group, Joe Richards, who soon collected a group of Americans at Defford. There the Fortresses of the American Eighth Bomber Command were fitted and tested with our H_2S and returned to their operational base at Alconbury. Toward the end of September the Fortresses used H_2S in a bombing raid on Emden. Meanwhile the early American opposition to H_2S had vanished and the U.S. Army Air Force had demanded that the Radiation Laboratory should produce twenty of their 3-centimeter versions of H_2S (known in America as H_2X). On 3 November 1943 the U.S. Air Force used this H_2X in a daylight bombing raid on Wilhelmshaven. The American system was simply an Americanized version of H_2S working on X-band (3-centimeter wavelength); our own 3-centimeter version was called H_2S Mark III. The advent of the American H_2X fitted in the Flying Fortresses released us from an imposed burden, and we quickly forgot the ironic nature of this association.

The Last Years of the War

During the early months of 1944 the 3-centimeter version of H_2S in the Pathfinder Lancasters continued to help Bomber Command in its task of striking at the industrial targets deep in Germany. The 10-centimeter version was being used by the main force, but there were many problems. It was evident that H_2S was difficult to use with the amount of the training given to the navigators. The defenses of Germany had attained new degrees of effectiveness, and the altitude of operation and the amount of evasive action increased to an extent that spoiled the H_2S picture so much that only skilled people could be expected to use it effectively under the stress of battle. Moreover, the results that people expected from H_2S had completely changed since its initiation. It was no longer a question of finding a built-up area but more of dropping markers and bombs around a precise aiming point in a given town—and a 50 percent zone of more than two miles was now considered hopeless.

We developed and introduced many changes to overcome the difficulties. Some were major, such as stabilizing the rotating scanner so that it remained rotating horizontally as the aircraft took evasive action. This and many other improvements did eventually materialize, but the delays became longer and longer. We had overloaded ourselves and the manufacturers. Matters got steadily worse and worse. The expectation of D-day turned people's minds to the end of the war, and the incidence of the flying-bomb attacks (the V-1)* put a stop to almost any worthwhile progress.

Meanwhile we had turned our attention to the achievement of even better resolution. In July 1944 we first flew a Lancaster using a 6-foot scanner (instead of the 3-foot scanner of the original H_2S). By this time equipment working on 1.25 centimeters became available, and in December 1944 we flew a system using a 6-foot scanner with remarkable results. Railways, runways, flooded fields, narrow streams, and bridges were clearly defined on the H_2S screen. Now the army generals became interested, and demonstrations were given of the detection of tanks. In the spring of 1945 a special wing was set up in Three Group of Bomber Command to carry out tactical work for the army using the Lancasters equipped with the 6-foot scanners working on 1.25 centimeters. Before the training of the operators was accomplished, the war in Europe had ended, and for security reasons an embargo was placed on the use of the 1.25-centimeter H_2S. So this highest definition of all our H_2S equipment never became operational—in either the 1.25- or the 3-centimeter version.†

The H_2S devices used operationally over Germany and against the U-boats in 1943 were imperfect systems. In time of peace they would be regarded as merely experimental. However, the early H_2S systems met a crucial operational need where immediate measures were prized above eventual elegant solutions. After the war there appeared elegant versions of our wartime equipment, but many became obsolete before there was ever an operational

---

*The V-1 (the "flying" or "buzz" bomb) attacks began on 12 June 1944. The launching sites were captured on 5 September 1944, by which time more than eight thousand had been launched. More than three-quarters of them were reported over England. Half of these were destroyed, but the remainder killed over six thousand people.

†Before the embargo on the use of the 1.25-centimeter version was made, six Mosquito aircraft fitted with this equipment and a 3-foot scanner carried out three successful marking raids at low level.

need. It was a vital aspect of the Allied war effort that the intimate association of the scientists with the operational staff made it possible for the scientists to recognize the urgency of an operational requirement and for the operational staff to accept with enthusiasm an immediate though imperfect solution that helped them in the task of defeating the enemy.

Two months after the end of the war in Germany, I was released from my wartime duties and returned to Manchester. In the weeks of relative quiescence before my departure, I had time to collect documents; the volume of these and my accompanying notes have made it possible for me to write these chapters about the war. The last phase had been increasingly acrimonious. Jealousies had developed among the groups of Bomber Command, and there were endless and tiresome arguments about the many versions of the H_2S that we had devised. I had become utterly exhausted after six years of that intense pressure. In a photograph of my group taken in February 1945 I am scarcely recognizable as a pale and wan figure. In a brief note I was given my official release as from 15 July, thankful to leave others to worry and argue about which versions of H_2S should go to the Pacific.

Only in my later years, and especially while writing about these events more than four decades later, have I been worried about the morality of my wartime tasks. As I left TRE, Air Marshal Sir Victor Tait wrote from the Air Ministry, "The radar equipments which you developed and saw into operational use in the RAF were one of the main factors in making the Air Force so effective in the destruction of the German armed forces," and Rowe wrote me that if I never did another stroke of work in my life I would have justified my existence. In retrospect these are worrying thoughts, and I am now increasingly thankful that the device for which I reluctantly accepted responsibility at the end of 1941 helped save as many lives as it destroyed.

7

The Transition from War
to Peace—Jodrell Bank

In the early afternoon of 8 May 1945 Prime Minister Churchill
rose from his place in the House of Commons to announce that
Germany had signed an act of unconditional surrender. At that
moment I was with Group Captain Dudley Saward, the chief signals officer of Bomber Command, with whom I had collaborated
closely in the preceding years to help Bomber Command locate its
targets. Through the frustrations and operational excitements of
those years, we had formed a close friendship, but our ways were
about to part and I saw little of Saward until forty years later he
wrote my biography. Only then did I learn our exact location on
that May day in 1945. I had remembered a rather remote country
house in a large estate and camouflaged huts in which miniaturized radio equipment was being assembled for the use of agents.
It was Whaddon Hall, near Bletchley, in Buckinghamshire, the
training and communications center for our agents operating in
enemy territory. Such was the secrecy surrounding this place that
I had not known that some of my equipment made for Bomber
Command had been used to locate the dropping-off and pick-up
position of the agents in enemy territory. Saward had taken me
there so that the brigadier in charge and his officers could thank
me for the assistance we had unknowingly given.

The Months of Uncertainty

I was in my thirty-second year, and it was nearly six years since I had left the research laboratory in Manchester. Circumstances had injected me into the thick of the political and military pressure to apply what already existed for operational ends. During the critical years only the question of ceaseless work for survival had reality, and the research laboratories of peacetime appeared as a mirage that had never been and could never be again. Then, after D-day, and particularly as the Allied armies swept across Europe in the early months of 1945, the pressures on us slackened and, like many of my prewar university colleagues, I found that the urge to return to scientific research became predominant.

After D-day, when victory rather than defeat and occupation seemed probable, preparations had begun in Britain to restore scientists to the universities and to industry. Many of my associates had simply been given leave of absence without pay by the universities in 1939, but my case was different. I had no university post to which I could return. My three-year assistant lectureship in Manchester had ceased in 1939, and Blackett had obtained a fellowship for me to continue with the proposed research on the Pic du Midi. I need not have worried on that point, since Blackett himself had taken action and persuaded the university authorities to offer me the post of lecturer in physics. The registrar of the university wrote on 26 January 1945 offering me this post, to commence on 29 September 1945, but I did not immediately accept this offer.

Although the urge to return to real scientific research and free myself from the restrictions of the service establishment existed, I faced a number of problems. P. I. Dee had been my superior officer for the last six years, and we had formed an intimate partnership. He had accepted the offer of the post of professor of natural philosophy in the University of Glasgow—Kelvin's former post—and some of my colleagues assumed that I would, and should, follow him to Glasgow. However, my real allegiance was to Blackett. It was with him, as much as in 1936, that I wanted to work. The doubt was whether he would return to Manchester. In 1945 he was the director of naval operational research and had the highest contacts in the government. At the same time pressure was exerted on me to remain in TRE. That was the last thing I intended to do, but I was less certain about offers from industry. There were two from separate divisions of Imperial Chemical Industries. I visited one of these, and was pressed to accept the post offered, but was so appalled by the conditions and the state of their supposed research laboratories that I left the scene as quickly as possible. Another I considered much more seriously. It was from the Christie Hospital and Holt Radium Institute, in Manchester, which wanted a physicist to pursue the new forms of treatment being opened up by the development of particle accelerators. Since the person appointed would also be an honorary lecturer on the staff of the university physics department, this had many attractions. But I was not a nuclear physicist and could not imagine myself being in Manchester, yet not working on Blackett's staff.

Another major problem was of my own confidence. The urge to return to research and rid myself of the encumbrance of a wartime establishment was accompanied by the self-doubt as to whether after a six years' absence I could really do this. To reeducate oneself after such an absence from academic life, from libraries, and from scientific journals seemed a formidable prospect. I sought advice from A. M. Tyndall, professor of physics in Bristol, where I was an undergraduate and graduate student, and had endless talks with Dee. The burden of their advice was to return to Manchester with Blackett. From Blackett at the Admiralty a letter indicative of his own intentions to return settled my fate. He wrote, "I am beginning to champ at the bit to get started. The radio detection of big showers and the distance to the moon must be fully thought out." So it happened that weeks before VE-day I had signed the conditions of appointment for the post of a lecturer in physics in Manchester, to begin on 29 September, provided I could obtain release from the wartime post.* The salary offered was £550 per annum, less than half my salary at TRE, where I had progressed to be a principal scientific officer, and far below that of the industrial posts. The salary issue had no bearing on my decision to return. The built-in passion for research, for the

---

*I was officially released on 19 July 1945.

acquisition of new knowledge, had survived the six years of conflict. The only real issue at stake was my own ability to make the transition.

The Return to Manchester

My resolve was soon to be put to the test. I expected to return to the world of 1939. In the stresses of war the transition to peace had a dreamlike quality. At last it seemed one would be able to switch on the lights, remove the blackouts, and enjoy the comforts and freedom of the prewar world. But, like everyone else who returned to the peacetime existence of 1945, I was soon to discover that six years of war had changed the nation and the world. Churchill had been deposed, and a Labour government came to power. Life grew harder, not easier. The food rationing of wartime was intensified; even bread was now rationed, a circumstance that had not occurred during the war. Everything was in short supply. Clothes rationing continued; petrol was rationed and hard to get. Nearly all construction materials were strictly reserved for essential national needs. Those of us who wanted steel or timber could not get it. A system of licensing was introduced, and the dead hand of the bureaucrats began to stifle effort and initiative. The very business of living demanded efforts of will and time without precedent.

However, by August of 1945 we had purchased a pleasant house and garden in an outlying suburb of Manchester. Once more we were established as a family, not far from the house we had left in 1939, and from that stable and happy background I was soon able to journey to the university to begin the restoration of my academic career. Manchester had been bombed but by no means devastated. In particular the dark, grime-encrusted Victorian buildings of the university and the surrounding dreary slums had survived unscathed. The physics department was almost deserted. The few members of staff who were there greeted me without enthusiasm. I had to realize that they, too, had lived through the war. Somehow the fabric of academic life had to be maintained. There were still students to be taught. It had been their duty to keep the university going academically and to preserve it from destruction by doing their duties as air raid wardens during the long, cold winter nights. If I had expected to be received back with admiration, I was gravely mistaken.

I retreated to the university staff room, where no one knew me and I knew no one. It was a bad beginning for a man who had recently been in the company of the commanders in chief. I became aware that the transition to research was not the only problem that faced me. Almost precisely six years earlier I had switched off the power to my cloud chamber. I found it exactly as I had left it except that it was now covered in thick dust. At least I could make a start on that and began the process of cleaning and restoration. My own hands again—no buttons to press for assistants, no secretary to summon for dictation. After a few days I wanted a new part—costing a few pounds only. But who would pay? The steward said he had no money for things like that. I was bewildered. I had been the man who could summon a heavy bomber simply with a telephone call. I had not heard about the cost of devices for six years. The problem had never been could we get the money but only could we do it? would it work in time? After a few weeks I began to assess the real problems of the transition. Six years earlier I had had to learn that dealing with people and the organizations was a major part of the problem. Now only myself and the apparatus were the problem. The transition was in reverse, but the glamour, the excitement, and the fear of 1939 were replaced by the loneliness and emptiness of the laboratory and by a cloud chamber that would not work, because a new part was needed.

Then suddenly Blackett appeared. Yes, he had resigned as director of naval operational research and was assuming his university duties again—and why was I messing around with that cloud chamber? Surely we had agreed, he said, that I would get radio echoes from cosmic-ray showers. Had I forgotten? Yes, I had completely forgotten the glamorous ideas of six years earlier. Blackett guided me to the copy of the paper that had appeared under our joint authorship five years before in the *Proceedings of the Royal Society* and instructed me firmly to get on with the job. I looked at the simple formula to see what was needed. It was all straightforward—high transmitter power, high aerial gain, maximum receiver sensitivity, and long wavelength. I began to feel at home again. This was exactly the gear I had been dealing with for six years—except the wavelength. I had spent the war pushing the wavelength down and down, and that centimetric equipment would be no good for the cosmic-ray ionization experiment. A radar CH station, like the one at Staxton Wold on which I had seen the transient echoes, would be ideal, but I could scarcely transport that to Manchester. And so the first peacetime advantage of the multiplicity of wartime contacts came into play. Not long before, I had visited the Army Operational Research Group at Byfleet and had met J. S. Hey, a prewar Manchester graduate. He was concerned with the army gun-laying radar—that is, the mobile radar for detecting enemy aircraft and directing the anti-aircraft guns. That worked on a wavelength of 4.2 meters, had a directional Yagi aerial system and a powerful transmitter, and was, furthermore, mobile. My hope that, with the cessation of hostilities, Hey might be able to arrange a loan of one of these systems was soon justified. In this way on a September day a large army convoy arrived in the quadrangle outside the physics department. The truck and the army personnel soon departed, leaving behind three large trailers—one containing the transmitter, another the receiver and cathode-ray tube display, and a third the diesel generator to supply the power. To the wartime university staff, who had handled no more than a bucket of sand and a long-handled spade, this caused utter consternation, and my ejection from the university together with the trailers was avoided only by Blackett's intervention with the vice-chancellor.

The army technicians had kindly erected the Yagi aerial arrays in their normal positions on the roof of the transmitter and receiver cabins, and soon I had the radar working. To my dismay the cathode-ray tube display, on which I had hoped to detect occasional transient echoes, was obliterated with noise the origin of which I quickly identified as the electric trams running past the university on a road only twenty yards from my receiver. The electrical supply to Manchester and the university was still an ancient direct-current system, and the sparks from the overhead lines of the tramway systems caused the intense interference on the radar receiver. I discussed this with Blackett, who suggested that we try again in the early hours of a Sunday morning, when one might expect the interference to be very much less. We tried, and Blackett saw for himself that the interference was still far too severe for any serious investigations to be made on that site. He agreed that I should try to find a site away from this electrical interference in order to search for the transient echoes.

Jodrell Bank

When I had first walked into the staff room of the university a few weeks earlier, I was, and felt like, a stranger. After that early Sunday morning test with Blackett my luck changed. Someone recognized me and waved across the room. It was R. A. Rainford, then the deputy bursar of the university. At last here was someone who seemed pleased to have me in the university again. He had heard about the trailers outside the physics department and was anxious to know the story. I told him the plans and our troubles and asked if by chance the university owned any land outside the city boundaries to which I could transport the trailers for a short experiment. After some thought Rainford suggested that I look at a small piece of land near Lymm that the university owned. Lymm was then a small community on the outskirts of the city of Manchester, and following Rainford's directions I drove there to see if it offered any hope for my purpose. It did not. The afternoon was wet, high-voltage grid lines traversed the field, and from a distance I could hear the sizzling noise of electrical discharges across damp insulators that would have obliterated the cathode-ray tube display with noise even more thoroughly than did the electric trams running past the university.

Somewhat disconsolately the next day I again approached Rainford. We were in the room of the staff house, where the university staff assembled after their luncheon to drink coffee and gossip before starting their afternoon lectures. Temporarily withdrawing from his mouth the pipe he habitually smoked and pointing its stem across the room, Rainford directed me to a bearded man holding a glass of beer in his hand. "Try him—his

111

name is Frederick Sansome, a botanist—and he runs a few acres of trial grounds for the botanical department quite a few miles south of the city." If Rainford thought that was the end of the matter as far as he was concerned, he was never more mistaken.

The botanist, Sansome, could well have been the type of possessive member of the university staff, too jealous of his work to welcome intruders. If he had been, and if chance had not placed him in the line of sight of Rainford at that moment, Jodrell Bank might well have remained an unknown strip of land on which the botanists tested the effects of water on the growth of carrots. But Sansome was not that type. True, he was a professional botanist, but professionals are often more interested in their hobbies than in their daily work. By an extraordinary coincidence Sansome was an amateur radio enthusiast, and within moments of my approach to him he was asking penetrating questions about the radar system in my trailers. There could not possibly have been a more felicitous beginning to my request, first to look at his place and then, if suitable, to take my trailers there. Sansome's only doubts—that it was remote from the city and without supply of electricity—were precisely the reasons that filled me with hope.

When Sansome gave me directions to this place I had no idea of its location. With his sketch of the route before me, I drove south from the university and eventually left the last of the city's dormitory communities behind me. After ten miles I saw the small garage that Sansome had described as the nearest landmark to his experimental grounds. Today, that same small garage still exists, although the construction of a major motorway nearby has injected a good deal of traffic along the road. On that day in late November 1945 the place was remote and deserted. Petrol rationing was still in force, and there was no traffic. Sansome's place was on the opposite side of the road. Continuing south I found the entrance—rough gates made of wire netting by the side of which stood a small notice board: "University of Manchester, Jodrell Bank Experimental Grounds." I drove in along a rough dirt track through a small grove of trees. Not more than twenty yards from the entrance stood the only two buildings—small wooden huts in one of which I found the friendly gardeners Alf Dean and Frank Foden warming themselves before a coke stove.

Sansome was certainly right to warn me that it was remote. Although in the years to come, bemoaning the encroachments of traffic and modern facilities, I was to search for remoter places, on that day, its isolation enhanced by the November fog, and the nearest electricity supply lines many miles away, the place was ideal for my purpose. Nearly a mile from the center of the small community of Lower Withington, this was Jodrell Bank. Soon I was to discover its history. *Bank,* because the land stood on the edge of a small escarpment leading down to the Red Lion Brook. *Jodrell,* because of its historical connection with the ancient family of Jodrells. William Jauderell, who fought as an archer with the Black Prince at Poitiers in the fourteenth century, was given a pass by the prince to return to England because his home had been destroyed by fire. The pass is still in the possession of his descendants. One of these descendants married a rich heiress from Twemlow, the next community several miles south of Lower Withington, and the Jodrell family, as the name became spelled over the years, owned a large estate in this region of the county of Cheshire. In the nineteenth century one branch of the Jodrell family became sponsors of scientific activity. In the University of London they endowed the Jodrell Chair of Physiology and founded the Jodrell Laboratory at Kew Gardens, near London. In the Frick Collection, in New York, a Gainsborough portrait of a member of this family graces the walls.

Eventually the estates in Cheshire were split into many small farms. In 1939, shortly before the outbreak of the war, the university purchased eleven acres of land from one of these farmers, Ted Moston, for use by the botany department. There were three fields, and the only one that had so far been used by the botanists was where I ended on that day in November 1945, enjoying the warmth of the stove with Alf and Frank, thinking only of the experiment I wanted to do in the next few weeks, without any vision of the future. My immediate problem was to get permission to bring the trailers to this place—and to tow them there. Sansome, who was to take a post in Nigeria sometime later, was soon to vanish from my life, but at that moment he was a critically useful person. He readily agreed that I could site the receiver trailer near the wooden huts. The transmitter and diesel generator

113

I wanted to place at least a hundred yards away from the receiver. As far as he was concerned, there was no problem. The hard dirt track continued as a soft muddy strip to a manure heap. That was ideal for me, provided I could get the heavy transmitter cabin there. With these arrangements amicably agreed with Sansome, he advised me to get a more authoritative permission than he could give to move the trailers to this place. My world had indeed changed. Five years earlier I had commandeered an entire airfield by a single telephone call. Now there was no telephone and, as far as I could see, only a vague authority who might give me permission to move three trailers from one place to another. The transition to peacetime life and research was beginning to appear a difficult and frustrating affair.

I have often wondered later in life what the consequences would have been had I been much older than my thirty-two years, without the experience of the war years and without the favorable ambience of the establishments that I had forsaken but whose staff were still very willing to extend a helping hand. On that gray and foggy winter's day I might well have returned to the comforts of fireside and home. Instead, the passionate desire to do this experiment led me back to Manchester and to Blackett. I told him that I had found an ideal site, that Sansome was willing and cooperative, but that some higher authority was needed before I could move the trailers. Blackett had faced a similar, though lesser, problem when he brought his large cosmic-ray cloud chamber and magnet system to the university in 1938. That was still by far the largest piece of scientific apparatus that had ever crossed the university threshold. But my trailers were far bigger and seemed to pose a mysterious threat to the sanctity of university life. Indeed, fifteen years later the outcome of my activities was to do precisely that, but at that moment it seemed to Blackett and myself that the issue was a purely practical one, involving no consequences for the future other than a limited exercise in scientific research. He promised to see the vice-chancellor. Although I did not then know the vice-chancellor, I was soon to learn that it was my immense good fortune that he was Sir John Stopford, a distinguished medical scientist and a fellow of the Royal Society. Although he was now engrossed in the heavy burdens of an administration trying to resuscitate a university in the postwar era, his outlook and attitude remained that of a scientist. Assured by Blackett that my proposed move to the botanical grounds would be for a very short period and that this would not interfere with my university teaching duties, he gave me permission to move the radar trailers to Jodrell Bank for a period of two weeks.

The next problem was to dismantle the Yagi aerials from the top of the transmitter and receiver cabins and tow the three trailers twenty-five miles south to Jodrell Bank. I begged assistance from Hey, who readily sent up an army team of technicians and trucks from Surrey. Exactly how the cost of this extensive operation was covered I never inquired or discovered—they all seemed delighted to have a job to do, and I imagine that, like many other service people, they were simply waiting to be demobilized. In any event, in early December the move took place, uneventfully until we attempted to place the transmitter and diesel near the manure heap along that muddy track at Jodrell. I have on record a photograph I took on the first day at Jodrell Bank, showing the transmitter and diesel trailers sunk to their axles in that sticky mud.

The First Days at Jodrell Bank

After two days' work the army contingent manipulated the transmitter and trailer onto the level ground and, after reerecting the Yagi aerials on the roof of the two cabins, departed. I was alone, it was 12 December, and since darkness was falling and I had no power or light I could do nothing until the next day. When I arrived on the morning of the thirteenth, the first job was to start the diesel. It was a heavy three-cylinder device, and the procedure for starting was to grasp the wooden handle with both hands and wind it round and round until the engine sputtered into life. I was then quite strong and had previously just had enough strength to start this machine. But now there was no sign of life in the engine. Exhausted, I sought the help of Alf and Frank, the two gardeners, but even with their greater strength no life came from the engine. We retired to the warmth of the coke stove and our sandwiches to discuss the problem. Their wise answer was to seek the help of the farmer Moston, who they said knew all about diesels because he had recently purchased a tractor. Moston was only too happy to gain a closer acquaintance with the trailers that had appeared on his former farmland, but he, too, failed to start the machine. However, like many of the country farmers of that postwar era, he had learned to depend on his own skills and initiative. By nightfall he had much of the machine in pieces. The next day he discovered the trouble. The prevailing mud and rain had turned to ice, and so had water that had leaked into the fuel pipes. With a show of immense pride Moston cleared the blockage and soon had the diesel generating the first electrical power that had ever been produced in that quiet Cheshire countryside.

Now it was my turn to try a start on my postwar scientific career. Late afternoon on 14 December 1945. The notebook still exists—power generated, transmitter switched on according to army instructions, can draw sparks with my fingers from the transmission lines to the aerials, so radar pulses being transmitted into space, retire to receiver cabin, switch on receiver, and anxiously gaze at cathode-ray tube. Several short-lived echoes at various ranges seen on tube. Scientific research at last. It was dark and late, but I was elated and excited. Six years of longing and several months of doubt and frustration were behind me. Again I was a real scientist able to study the universe with a new technique.

But was this really the case? The next day and the next the transient echoes were there on the tube—a few every hour. I proudly told Blackett that the radar was working, free of interference, that there were present the transient echoes that I had seen on the defense radar six years earlier on the first day of the war. By this time Blackett had already collected several members of his prewar staff, and many new faces were appearing every day in the laboratory. Talk about science and especially about cosmic rays now filled the laboratory, and once more we were able to study the scientific journals. From these I suddenly learned why Blackett was so enthusiastic about my idea of studying the upper range of the cosmic-ray spectrum by using radar to detect the most energetic showers. In the early days of the war he had delivered the Guthrie Lecture to the Physical Society of London. He had there drawn attention to the remarkable fact that the energy spectrum of the cosmic-ray particles incident on the atmosphere followed an inverse-square law over a range of energies of 10^{10}. He thought this must have some deep cosmological significance such as that of the Hubble law for the expansion of the universe. At that time the upper range of the spectrum was known only to about 10^{16} electron volts (eV), and evidently a matter that might be of deep cosmological significance was the question of how far this inverse-square law applied at even higher energies.* Although Blackett's own researches on cosmic rays at that time were beginning to develop in many directions, his enthusiasm for my idea of the radar detection of the showers was therefore, understandably, as great as my own, and he wanted to see the echoes himself. He was fascinated by the occasional random appearance of these short-lived echoes on the cathode-ray tube but soon began to doubt whether they could be associated with the cosmic-ray ionization—we had expected one or two per day on this equipment, not several per hour.

The V-1 and V-2 Echoes

Blackett thought that Hey, after his years of wartime experience with this equipment, must have seen these echoes and suggested

---

*Today, more than forty years later, the problem remains substantially unsolved. A number of showers have been recorded implying a primary energy of 10^{19} eV, and there is evidence for a change in the spectral law. At lower energies the photographic emulsion techniques have revealed that the primary particles from space consist of about 90 percent protons, with the remainder mainly alpha particles. At the high-energy end of the spectrum the primary particles are believed to consist only of protons, probably generated in supernovae or pulsars. This extreme upper-energy end of the spectrum remains of great scientific interest. Blackett's intuition that the form of the spectral law might be of cosmological significance may well be justified. In theory there are fundamental processes involving the collision of very high-energy protons with the microwave background radiation (the relic radiation from the early state of the universe) that would degrade the energy of the protons so that protons of energy greater than 5×10^{19} eV may not exist. The determination of the spectrum of the upper energy range, particularly to discover whether there is a final cutoff in energy of the primary cosmic rays between 10^{19} and 10^{20} eV, therefore remains a problem of cosmological importance.

that I discuss this with him. Early in 1946 I did so, at which time Hey told me the following remarkable story. In June 1944 the Germans began to bombard London and the south of England with the V-1 missile (the buzz bomb). These pilotless planes were launched from sites on the north coast of France, aimed in the general direction of London; when the fuel was exhausted, they fell to earth and their heavy bomb load exploded. Hey had been instructed to modify the 4-meter anti-aircraft gun-laying radars so far used to direct the anti-aircraft guns to high-flying bombers so that these low-flying V-1s could be detected and attacked by guns and fighter aircraft before they reached London. Hey did this simply by lowering the Yagi arrays in such a way that the aerial beam was horizontal. Although this led to the early location and destruction of many V-1s, the Germans had launched more than eight thousand of these devices against London by the time the Allies had captured the launching sites in early September. Since in this brief period the V-1s had killed over six thousand people and injured another eighteen thousand, there was great relief at the cessation of the attacks. Unfortunately, this relief was short-lived.

On 9 September 1944 the Germans launched the first of the V-2 rockets against London. This was a ballistic rocket against which there was no effective countermeasure, since the rocket attained an altitude of 60 miles and fell on London 200 miles from the launching sites at a speed of 3,500 miles per hour, five minutes after lift-off. Although there was no means of intercepting and destroying this rocket, at least civilian casualties might be avoided if it could be detected during its ballistic trajectory, so that the sirens could give the population a few minutes warning to take cover. Hey was once more asked to investigate this possibility of detection by using the 4-meter anti-aircraft radars. He did this by adjusting the Yagi aerial arrays on the roof of the transmitter and receiver cabins to an elevation of about 60 degrees and directing them to the general position of the launching sites. Before the Allies had captured all the launching sites by the end of March 1945, 1,115 of these V-2 weapons had fallen on London and southern England. They had killed 2,754 people and injured another 6,523. There would have been many more fatalities had not the radars been able to detect the echo from the rocket in flight and so enable a brief warning to be given. Unfortunately considerable confusion occurred because the radars often detected echoes and gave warnings when no V-2 materialized—and furthermore the Allied agents confirmed that no V-2s had been launched at the times when the radar echoes led to these false warnings.

Hey had been given the task of investigating these peculiar occurrences, and when I spoke to him early in 1946 he handed me a copy of a secret report that contained the results of his analysis. The conclusion was that many of the short-lived echoes that had caused false warnings to be given of the approach of a V-2 rocket originated in the ionospheric regions at a height of about one hundred kilometers. Some, at least, appeared to be associated with the entry of meteors into the earth's atmosphere. In the report Hey referred to some prewar work by ionospheric investigators, specifically to that of J. P. Schafer and W. M. Goodall and A. M. Skellett in America, who claimed to have identified certain types of sporadic echoes on the long-wave ionospheric sounding equipment with the ionization caused by meteors. Hey told me that since this report was issued he had been able to study the problem again after the end of hostilities and that during the autumn months of 1945 he had been able to confirm that the appearance of a meteor streak in the sky often coincided with the occurrence of a transient echo on the radar equipment. However, the majority of the echoes had no obvious association with the sighting of a meteor streak in the sky, and their origin was not known.

The results of my conversations with Hey were therefore not conclusive as regards the echoes I was observing at Jodrell Bank. According to Hey some must be from meteor ionization, but it was not possible to exclude that others might be the effect for which I was searching, that is, the radar reflections from the ionization caused by large cosmic-ray air showers. Clearly the immediate task was to find out more about meteors and to understand their effects in order to identify any echoes that might arise from cosmic-ray showers. I was not an astronomer and knew little about meteors. There were very few astronomical books in the university library. Today, when so many popular and professional astronomical works are published, it is difficult to recall that immediate postwar situation. The available texts made little reference to meteors, and the few professional astronomers at that time working in England did not seem able to offer any help. I got the impression that the subject was somewhat beneath their dignity and that meteors were not objects on which they used their telescopes.

The Amateurs—Manning Prentice

It was now the spring of 1946 and, while I was still puzzling how to obtain enough information about meteors to help me decide which types of echoes on the radar could be associated with them, a marvelous stroke of good fortune occurred. Nicolai Herlofson had escaped from Norway during the German occupation and had then worked in England as a meteorologist. Blackett thought that, as a meteorologist, Herlofson might also know something about meteors. I must confess that in later years I often wondered why Blackett thought there would be a connection between meteorology and meteors. However, this turned out to be an inspired move since, although Herlofson did not know much more about meteors than we did, he did provide a vital piece of information. His response to my query was that it was a waste of time to seek help from the professionals because it was the amateur astronomers, especially in Britain and America, who were the principal investigators of meteors. Furthermore, Herlofson knew that in England there existed an active group of amateur astronomers organized within the British Astronomical Association into specialist groups responsible for regular observational work on all the heavenly objects to which amateurs using small telescopes could make useful contributions. These specialist sections dealt, for example, with the moon, the various planets, comets, and of course meteors. I was soon in touch with the association and particularly with the director of the meteor section, J. P. M. Prentice. That contact was to be of the greatest importance to my postwar career. It was Manning Prentice, a solicitor working in the county of Suffolk, in the east of England, who changed my career from that of a physicist to that of an astronomer. Prentice reacted with instant enthusiasm to my approach. He explained that although meteors could be seen at any time of the year they were sporadic and that the most important events occurred regularly at certain times of the year. These were the shower meteors when the visible hourly rate rose from a few per hour to fifty or a hundred per hour for several successive nights. He mentioned that many of these known meteor showers occurred when the earth crossed the orbit of a comet but that there was uncertainty about the origin of the sporadic meteors, some authorities maintaining they were of interstellar origin.

When I explained to Prentice the nature of my problem, he immediately responded that I should use the radar equipment throughout the period of one of these major showers in order to discover whether the echo rate was related to the known meteor activity. It was then late spring, and Prentice said the next major shower would be the Perseids in August. Furthermore, he offered to come to Jodrell for that event, saying that he had in any case planned to take a holiday from his office at that time in order to concentrate on the observation of the Perseids and that he could observe them as well from Jodrell as from his own garden.

Manning Prentice turned out to be a remarkable man. His devotion to meteor astronomy was extraordinary. During all the major showers, he would observe throughout the night without interrupting his professional work as a solicitor. He had a team of observers scattered throughout the country from whom he would collate the observations and make regular reports, published in the journal of the association. He had himself published a considerable number of papers on meteors and on 13 December 1934—while observing one of the major showers, the Geminids—had discovered a new nova, Nova DQ Herculis.

This discovery of Nova Herculis and Prentice's prompt action, which enabled the spectra of the early stages of the outburst to be obtained, was a major astronomical event. *The Times* of London carried a detailed description of the discovery on 14 December 1934, and on 17 December the leader writer paid tribute to Prentice:

Astronomer by Chance

Astronomers are already paying their homage, and from countless observatories telescopes now point to "Nova Herculis". . . . But let it not be forgotten that it was a Stowmarket solicitor, Mr. J. P. M. Prentice, who first observed that a star in Herculis had run amok. . . . The honours of astronomy not infrequently go to the amateur. . . .

Prentice had never received any training in astronomy, yet he had acquired the deep knowledge of the sky that enabled him instantly to recognize that "there was something wrong" with the appearance of a constellation. His curiosity about the heavens was far greater than that of many professional astronomers.

All of this I learned much later. At that moment in the spring of 1946 Prentice said that he would arrive at Jodrell one day in late July, and he did so—in an open car whose backseat was loaded with a deck chair, a flying suit, flashlights, a large celestial globe, various charts, and other miscellany. He explained that his technique of observing was to lie almost flat in the deck chair in the open. Meteors occurred without warning, and the visible trail disappeared in a fraction of a second. In order to get accurate data, Prentice had developed the idea of aligning a piece of string along the trail so that he could read off the beginning and end points relative to the stars at reasonable leisure. Using a dimmed flashlight so that his night vision was not impaired, he then quickly wrote down on a writing board strapped to his knee this and other basic data, such as the magnitude, time of flight, and time of occurrence. He would do this from dusk to dawn with occasional breaks, which he had found necessary to avoid visual fatigue. The flying suit was simply to keep him warm during the night. Later I was to be with Prentice when he observed in this manner throughout the whole of a winter's night with the deck chair on snow-covered ground. The only point that mattered to him was that the sky should be free from clouds and that the night should be moonless. It was truly an astonishing performance, carried out, I was to learn, against a background of associates who felt that his time would be better spent in his office or on church affairs. Prentice was a solicitor, domiciled in a farming community, trained to deal with tenancies, legacies, investments, and so on. At night he became a quite outstanding observational astronomer, and the turn of fate that led him through a legal training instead of to a scientific career was indeed responsible for a great loss for science.

August 1946—The Perseids

Fortunately during the Perseids of 1946 Prentice did not need his flying suit. Soon I found myself another chair and stayed by his side through those warm summer nights. He knew every star down to the fifth magnitude, which was the limit of our vision during those nights. I learned to recognize, as midnight approached, the stars of the Perseus constellation rising in the east, marveled that it was the rotation of the earth that moved the stars across the heavens, and suffered nostalgia as dawn broke and the stars slowly disappeared in the light of the morning sky. Prentice had learned his astronomy the hard way—by watching the skies—and that is how I learned mine during those August nights.

By the time Prentice arrived at Jodrell, there had already been substantial changes since my arrival there with three trailers the previous December. With the cessation of hostilities large amounts of equipment became surplus, and when I found out that much of this was being tipped into abandoned mine shafts, my acquisitive instincts were stimulated. Within a few months I had accumulated a large amount of this apparatus, which the university could never have purchased at its open-market value. The air marshal in the ministry with whom I had recently collaborated on military affairs agreed that I could take away almost anything I wanted if I would pay him £10, so that he could make an appropriate entry in the accounts. Blackett made a similar arrangement for some equipment I wanted from the Admiralty, and more of the army transmitters and receivers I obtained with the help of Hey. I estimated that, by the time Prentice arrived in late July, at current prices there was more than a million pounds worth of equipment in that field at Jodrell—equipment for which I had paid only a few pounds.

There had also been another important development. Blackett, enthusiastically expanding his physics department in many directions, realized that the radar techniques were opening a new era for research. He said that I must get some help and that if I knew people in TRE who might be searching for a university post, he would arrange to pay them. In that way I soon acquired a few helpers, the most important being J. A. Clegg, whose knowledge of aerial systems was to be of paramount importance. For some years during the war Clegg had taught the fundamentals of radar in the TRE school. He remained with me for only about five years, but during that period made fundamental contributions to our progress. Later he decided to work on measurements of rock magnetism, which Blackett was just developing, and his results were of great significance to the theory of continental drift. Another senior recruit was C. J. Banwell, who was soon to return to his native New Zealand, but who was to prove a valuable partner in these early developments. Among the equipment that I had borrowed from the Air Ministry was a large covered truck filled with the most sophisticated receiving and display equipment of the day. This was known in the services as a Park Royal—after the district near London where it was manufactured. My problem was to get this prime mover from its site near London to Jodrell. I had no experience of driving such heavy, cumbersome vehicles and was reluctant to try. Clegg arrived when I was pondering this difficulty and immediately claimed to have driven such vehicles as part of his wartime duties. In preparation for the arrival of the Park Royal we had, with the help of Moston's tractor, laid two tracks of stones over the quarter mile of soft ground from the road to this third field. Clegg drove the Park Royal vehicle safely the 180 miles from London to Jodrell and succeeded in negotiating this double track, but one hundred yards into the third field the truck sank to its axles into the soft ground, beside a small pond frequented by moorhens. We had no option but to leave it in that position. The engine was switched off and never restarted, and that vehicle's quite accidental position in the field was to determine much of the future layout of Jodrell—and the name of a nearby building that, years later, was to replace this original Park Royal.

It was in that stranded vehicle that we were working when Prentice arrived to observe the 1946 Perseid meteor shower. Our procedure was straightforward—and of a simplicity that would no doubt appall the computerized scientist of today. On arriving at dusk, we would first pour diesel fuel into the engine and hope that our combined efforts at winding the handle would start the engine and so produce the electrical power. Once the transmitter and receiver were switched on, I would take turns with Clegg and Banwell in the Park Royal. With eyes glued to the cathode-ray tube, we would record in a notebook the time, range, and duration of any echo that appeared on the screen. Prentice, in his deck chair, was established outside, as close as possible to the steps leading up to this receiver cabin. If we saw an echo, we would shout, and if he saw a visible meteor, so would he. In this fashion we were able to establish immediately and without any ambiguity whether there was a connection between the radar echoes and the visible meteor trail seen by Prentice.

We began these systematic all-night recordings on 29 July. At the beginning of each night's watch our echo rate would be low—two or three per hour. After midnight, as the constellation of Perseus rose in the east, our echo rate began to increase, and Prentice began to see the Perseid meteors, which appeared to emanate from a point in the constellation. We learned that this was the radiant point of the shower. The position of this point was named after the constellation in which it appeared to lie and was defined accurately by the celestial coordinates of that point. Of course, the meteors had no connection with the stars of the constellation other than one of nomenclature. Prentice explained that during these August nights every year the earth moved through the orbit of a comet known as Comet 1862 III, and the debris scattered along the orbit of the comet plunged into the earth's atmosphere at high velocity, was burned up at a height of about one hundred kilometers, and produced the long visible streak of the meteor—commonly called a shooting star. He told us that the computation of the orbits of these meteors by G. V. Schiaparelli in 1864 and 1866 established their coincidence with the orbit of the Comet 1862 III and that this was the first occasion on which such a connection had been determined.

Every night our echo rate would reach a maximum when this radiant point was at its highest in the sky—due south on the meridian. As the radiant point set in the west at dawn, our echo rate decreased. This phenomenon occurred every night, but the echo rate—and the number of visible meteors seen by Prentice—increased night by night, to a maximum on 14 August, and then fell abruptly to a low value. The earth had emerged from the cometary debris, and our first astronomical observations were at an end.

The result was exciting. A large number of echoes were indeed coincident with the appearance of the visible meteor trail. There was no doubt that our radar pulses were being reflected from the ionized trail left by the meteor. However, it is rare that scientific observations are conclusive, and these certainly were not, because we had recorded many echoes when Prentice had not seen a meteor; conversely he had seen many meteors when we did not see an echo. On the whole it was our echoes of longest duration that coincided with the visible meteors, whereas the short-lived echoes had no obvious connection with the meteors seen by Prentice. Evidently there was still much to learn—and in particular we had still not excluded the possibility that some of the echoes were associated with the large cosmic-ray showers. The Perseid data formed the subject of my first scientific paper on astronomy. The detailed account was published under the authorship of Prentice, Banwell, and myself in the *Monthly Notices of the Royal Astronomical Society.*

The Great Giacobinid Meteor Shower of October 1946

Before Prentice sadly returned to his office in Suffolk, we naturally asked for his advice on how we could proceed to elucidate further the complexity of the echoes visible on our cathode-ray tube and their relation to the meteoric phenomenon. His advice was decisive. He said that during the night of 9–10 October the earth would cross the orbit of a comet known as the Giacobini-Zinner comet. Furthermore, it would do so quite close to the head of the comet, and he expected a brilliant but short-lived display of meteors. This comet had been discovered by M. Giacobini in

1900 and rediscovered by E. Zinner in 1913. In 1926 Prentice had discovered the meteor shower as the earth crossed the comet's orbit, and in 1933 on the same date in October there had been a tremendous display of meteors. Prentice bade us farewell with the sad news that he would not be able to join us for this Giacobinid shower—he had used up his holiday entitlement on the Perseids, but he would be observing the Giacobinids from his home in Suffolk.

The aerials on the top of the army radar cabins were broad beam. The entire cabin could be rotated in azimuth (that is, horizontally); otherwise the direction of the aerial beam was fixed. When we had transferred our receiver to the Park Royal vehicle, the receiver aerial could no longer be rotated in this way and we had observed the Perseid meteors with a simple broad-beam array fixed in direction. There was already evidence that the meteor trails reflected the radar pulses specularly—that is, a strong echo was produced only when the meteor passed through the aerial beam at right angles to the axis of the beam. In fact, in our calculations for the cosmic-ray reflection experiment, Blackett and I had assumed that the reflection process could be treated simply as that which would occur from a long, thin column of electrons. The meteor trail ionization was evidently precisely this—a long, thin column of electrons—and the specular reflection to be expected implied that the maximum response would be that when the trail of electrons passed at right angles through the beam of the aerial. All of this reasoning was elementary physics, based on a well-known optical analogy, but the practical issue of how best to achieve the optimum arrangement for observing the meteors was a different matter.

The shower meteors, although apparently diverging from a point in the sky because of perspective effects, entered the atmosphere in parallel paths. Thus, if the reflection from the trails was specular, we needed an aerial that could be moved continuously both in elevation and in azimuth, with the axis of the beam always pointing at right angles to the paths of the meteors. Such a steerable aerial would also be an asset for the cosmic-ray observations, and in order to improve the sensitivity we wanted to build an array larger than those used with the army gun-laying radars. Clegg's wartime experience on aerial design proved an immense asset. He saw no problem in building an array many times larger than those we had inherited from the army. The problem was how to make it steerable. Whenever a practical problem like this arose, we immediately sought a solution from the equipment that had been used in the war and that was now lying around the country in considerable quantities. In this case the obvious answer was to use a large army searchlight. Although we did not want the searchlight mirror, the very strong rotating and elevating mechanism would be an ideal base on which to build the aerial.

Hostilities had ended a year ago, but for some peculiar reason the army said it still needed the searchlights. After a good deal of wrangling we eventually persuaded the army authorities to allow us to borrow a searchlight for a short period—without the mirror and provided we returned it in good condition. This was towed into Jodrell and deposited near the Park Royal vehicle. The wheels were removed so that the jacks attached to the solid base of the searchlight could be lowered to the ground. On this solid foundation Clegg quickly built a splendid array of Yagi aerials. In this way we first became involved in making substantial changes and additions to the radar equipment, which we had hitherto used substantially as it was during the war. There was a problem about the timber we needed for the support of the metal tubes that formed the aerial. The postwar shortage of timber was still such that there was strict rationing, and it would have been impossible to get a license to buy the material we required. We asked the gardeners Alf and Frank for advice, and they directed us to an ancient Elizabethan water mill in a neighboring village. We found the owner of this delightful spot, who was using a waterwheel to grind corn, drive a dynamo for his electrical supply, and operate a sawmill. If we expected a rebuff, we were mistaken. He had already heard about our equipment at Jodrell and was only too delighted to find the timber we wanted. No license required, and, yes, he would send us the bill. Forty years later I am still waiting for this bill, but the mill owner will have forgotten—he is happy that his grandson is now working as an electronics technician at Jodrell.

ASTRONOMER BY CHANCE

This "searchlight aerial" built by Clegg was an immense success. First used for our observation of the Giacobinid shower in October 1946, it survived for many years and became the aerial system for a variety of subsequent research programs. The army regularly reminded us that we should return the searchlight, but in the end even the army recognized that World War II searchlights would be of no use against high-flying bombers and missiles. Eventually this splendid aerial system was superseded, but the rusty remains of the searchlight base and the concrete trench that we built to carry the transmission lines remain to remind me of an epic night in our early history.

Indeed, the night of 9–10 October 1946 was unforgettable—a major turning point in our history. We started the diesel and our night watch at dusk. The night was brilliantly clear, and we eagerly watched the sky and our cathode-ray tube for signs of activity. There were none—just the normal two or three echoes per hour. Blackett wanted to see the promised spectacle but, disillusioned, said he had to be in London by the morning and left just before midnight. Banwell and Clegg were with me, and we were sadly beginning to feel that the expectations of a great shower were misplaced, when, just after midnight, the drama began. Suddenly and simultaneously the echoes appeared on our tube, and outside the sky seemed to be ablaze with streaks of light. One after another, with ever increasing rapidity, the trails of the meteors seemed to be emerging from some great source in the heavens and hurtling toward us with immense speed. By 3 A.M. we were in a frantic daze, unable to count either the meteors in the sky or the echoes on the tube. The earth was in the midst of the cometary debris, and although spellbound by the majestic sight, we retained our scientific instinct and at the peak of this display redirected the searchlight aerial so that the axis of the beam was pointing to the direction of the radiant—parallel to the paths of the meteors. The visual display was undiminished, but our echo rate instantly fell from a rate of over a thousand per hour to the two or three per hour that we had been observing before midnight. We returned the aerial beam to be at right angles to the meteor trails, and once more our great echo rate was sustained. Then, as suddenly as the display had begun, it ceased. By

129

6 A.M., with dawn slowly emerging, the sky became clear of meteors, and our echo rate dropped to its normal few per hour.

At 10 A.M. I was due to lecture on atomic physics to my undergraduate class in the university. In a state of elation I could think of nothing but meteors and, as many were to remind me in later years, talked to them of the Giacobinids and of the remarkable observations in that field at Jodrell. My preoccupation was understandable. Without question it was now clear that meteors gave radar echoes and that the short-lived echoes we had not been able to correlate with meteors in the Perseid shower were associated with meteors—probably with meteors too faint to be seen by the naked eye. Our turn of the aerial at the peak of the display had demonstrated that the primary reflection process was indeed specular—the echoes were obtained only when the meteors passed at right angles through the beam of the aerial— and that the radio waves were reflected from a long, thin column of electrons generated by the meteor as it burned up in the high atmosphere.

There was, however, a far more significant outcome than the purely scientific conclusions. We were asked to describe our work at the December meeting of the Royal Astronomical Society. At the beginning of the meeting on that Friday afternoon we were strangers to the astronomers, aliens infiltrating a privileged assembly. As we showed our slides, the mood changed. The fellows of the society began to grasp that this was a new astronomical technique, and by the end of the meeting we were part of the astronomical community.

8

Meteoric Phenomena

The meeting of the Royal Astronomical Society at which we described the results of our work on the Giacobinid meteor shower was on 13 December 1946. The meetings of that society are still held on the second Friday of the month, but today, with the large increase in the size of the astronomical community, they can no longer take place in the splendid but cramped quarters of the society's buildings in Burlington House, adjacent to the Royal Academy. In the early 1970s, although the society's administration and library remained in Burlington House, the monthly meetings were transferred to a nondescript government-sponsored lecture room a mile away. By modern standards, I suppose, everything is right about this room and almost everything was wrong about the room in which I spoke on that December afternoon. Then one entered through the magnificent portals of Burlington House, and the roar of Piccadilly disappeared behind one. The lecture room was small, and a hundred people would overflow from the tiered hard wooden benches to sit on the floor and stand in the doorway. There were to be many subsequent occasions on which I spoke in that room and, later, in the modern room with its rows of empty comfortable seats, but nothing can erase the memory of that first experience when I was not yet an astronomer.

I have once more taken from the library shelves volume 67 of the magazine *The Observatory*—the first issue of 1947, describing that meeting of 13 December 1946. H. H. Plaskett, the distinguished Oxford Savilian Professor of Astronomy, presided over the meeting as the president of the society. Promptly at 4:30 P.M. he called for the minutes of the previous meeting and after some official business asked T. G. Cowling to describe his work on the theory of sunspots. After various questions Plaskett then called upon "J. P. M. Prentice to give us an account of 'The Visual Observations of the Giacobinids 1946.' " Prentice proceeded to describe how the visual observations of his network of observers had been greatly hampered by bad weather and said that his account led "to the remarkable technical advances described in the two succeeding papers on the radio echo observations of the Giacobinids." He added, "It is obvious that this beautiful new technique represents a major advance in this field of observation; indeed we are like the first possessors of the telescope, unexpectedly armed with new powers of observation. With your permission Mr. President, I will ask Dr. A. C. B. Lovell to continue."

No one could have had a more gracious and encouraging introduction to face that packed room of establishment astronomers, and Prentice, as one of them, had skillfully prepared the ground for the accounts of a revolutionary technique. I was followed by Hey, who had also observed the Giacobinids, from the anti-aircraft defense site in Richmond Park. Sir Edward Appleton and many others spoke in the discussion until Plaskett rose and said, "In view of the lateness of the hour, I regret we must end the discussion at this point. I ask you to return your thanks to the speakers; they have presented to us an entirely new field of astronomical research."

The Great Daytime Meteor Showers

At that moment I doubt that Plaskett, Prentice, or we ourselves were aware of the scope of the new field of research or that within a few years the major contentious problems of meteors would be solved. The great hindrance to the study of meteors was that cloudy skies or a bright moon often interfered with the sys-

tematic observation of the phenomena. For the majority of astronomical observations the loss of a few nights because of a cloudy
sky is merely a nuisance, but the major meteor showers are short-
lived occurrences, and unfavorable conditions of sky or moon for
a few nights can obscure completely the annual record of showers such as the Perseids or the Geminids. It is easy to understand
why Prentice felt that with the radio echo technique, unhindered
by cloud or moon, we were like the first possessors of the telescope.

After the Giacobinids in October almost every night throughout
that autumn and winter would find us at Jodrell. We had lectures
to give during the daylight hours, and our lives became less regular than that of the night owl. The routine was simple but
arduous. Sometimes in ice and snow, often in rain, or under the
brilliance of a starlit sky, we would disturb the peace of the countryside by cranking the diesel generator until it sputtered into life
to give us electricity. With occasional readjustments of the direction of the searchlight aerial, we would then spend the night
hours gazing at the cathode-ray tube and writing in the notebook
the time, the range, the duration, and other details of any echo
that appeared.

It was a peculiarly obsessive occupation, never missing a
moment's watch in case the unexpected should happen. When
Plaskett closed that memorable meeting in London on 13 December, it was already half past six in the evening. I rushed to the
train for the journey that took three hours in those days, but not
to get home for a night's sleep. It was the anniversary of our first
switching on the radar in the trailers at Jodrell and near the maximum of the Geminid meteor shower. We had to spend that night,
like the preceding and following ones, at the trailer site, studying
the sky and recording the data by rudimentary methods that no
student would tolerate today. By these simple means we had
recorded every known meteor shower throughout the autumn and
winter of 1946–47. We knew that with this equipment our echo
rate bore a close relation to the number of meteors seen by a
visual observer under good sky conditions and that our determination of the positions of the radiant points of the showers agreed
well with the visually determined positions.

We had planned that our prolonged series of observations would cease in early May until the advent of the next series of known meteor showers, in late July and August. The last of the meteor showers known to the visual observers that we intended to observe was the η-Aquarids. The radiant point of this shower lay in the direction of the star η-Aquarii, and the observations early in May 1947 extended our watch into the dawn sky. The radiant did not cross our meridian until half past seven in the morning. It was inevitable then that we noticed a strange phenomenon. After the passage of the radiant of the η-Aquarids through our aerial beam, the echo rate did not decrease to the low value expected from the sporadic meteors. Soon it became obvious that this well-known shower, with its radiant near η-Aquarii, was not an isolated event but merely the beginning of a great belt of meteoric activity extending toward the sun and observable only in daylight.

Although we had noticed that during the summer months of 1946 our echo rate remained high, we did not grasp the significance of this until the autumn and winter studies of the known showers had established that our echo rate bore a close relation to the visually observed rate. Now we could also measure the radiant points of showers, and after a few days it became clear that the earth was moving through a vast region of debris in space, giving rise to spectacular meteor showers in the daytime, invisible and unknown to the human eye.

These phenomena developed with great rapidity, and soon it was evident that we had discovered not merely one new meteor shower but a whole series. By the end of June seven new centers of great activity had been delineated. This daytime activity continued throughout July and August, and a comparison with known meteor showers indicated that it was without precedent in extent and duration. Our constant watch of the cathode-ray tube in the trailer that had been transferred from night to day in May now had to cover night and day as the regular known nighttime shower of late July and August reappeared. Fortunately it was a time of great postwar expansion in British science, and Blackett had unhesitatingly handed over to me several of the students who were returning to the university to study for higher degrees.

Astronomer by Chance

Recently I discovered a photograph taken in late September of that year. It shows fourteen of us grouped in front of the searchlight aerial. Not only was the task of observing now divided, but the sheer labor of maintenance was shared by three technicians. Without premeditation our research was suddenly transformed into a new style, in which we needed people with technical skills to devise and maintain equipment as well as a team of observers. In retrospect I see that it was at this stage that the need to observe a natural phenomenon first introduced the problem of people and their administrative arrangements as an inescapable part of our research development. On that photograph I look happy, and so do the other thirteen. The photograph was probably taken when the great activity had at last ceased and we knew we had a major new astronomical technique at our command. However, the smile on my face, at least, hid several deep anxieties. The most profound of these, which eventually determined our future in a quite unexpected manner, I shall describe in the next chapter. The most pressing anxiety was of far less fundamental significance. I had continued to assemble a large miscellany of surplus equipment, larger diesels that were even harder to start, many trailers still in their camouflage paint, and an assortment of aerials scattered at various points over a muddy field that did not even belong to us. With good reason the locals described our operation as the fairground, and I had presented Blackett with a nasty problem. On the one hand he was delighted with our progress and day by day shared our mounting excitement.

On the other hand the temporary permission to bring our original trailers to these botanical grounds at Jodrell Bank had long since expired—the two-week permission had now extended to eleven months, and the vice-chancellor and the officers of the university were asking awkward questions about my absence from the university in the remote Cheshire field. The doubts whether people like myself could readjust themselves and again become satisfactory members of the university and able researchers still existed.

Neither I nor Blackett had authority to continue working in this way, a good hour's drive from the university campus where we should have been concentrating our teaching and research.

Understandably Blackett thought he must introduce to the vice-chancellor and the university authorities the idea that our activities at Jodrell had gone somewhat beyond those that could be carried out in a temporary trailer in a corner of the grounds of the botany department.

Although I viewed such an approach with some dismay and anxiety, I was beginning to feel the need for an official blessing to continue our work at Jodrell. In particular there was the urgent question of our trailers. They had been built to provide temporary transport and cover for wartime equipment; now, having admirably served that purpose, they were beginning to leak and disintegrate. The equipment in that field—borrowed, begged, or bought for a token sum—would have cost nearly a million pounds to replace. In a heavy thunderstorm in the midst of the excitement of observing the daytime meteor streams, the drip, drip of water on the transmitters and receivers was like sledgehammer blows on my head.

So it was that when Blackett announced that he had persuaded the vice-chancellor to visit Jodrell to see for himself what was taking place, I knew that our whole future was at stake. Vital matters of university policy were beginning to be involved—never before had the university sanctioned such a development remote from the campus. I have already had occasion to refer to our good fortune that the vice-chancellor, Sir John Stopford, was an eminent scientist, a person widely recognized for his unerring judgment. I knew that there would be no halfway compromise and that he was not a man to be impressed by the sight of a medley of decaying wartime trailers in a field or by talk of the monetary value of the equipment. It was June when Blackett decided to bring Sir John to visit us, and I begged him to do so early in the morning, when we could display the echoes from the great daytime meteor streams, and not in the afternoon, when the radiants would have set and the echo rate would be low.

How very different could that occasion have been! It could have been cold and wet, our diesel could have failed, or our equipment could have broken down—and there would have been no future for Jodrell. On the contrary, the mists of the early morning had disappeared to reveal the beauty of a midsummer morning that England alone can produce. The peace of the countryside was disturbed only by the sound of our generator—and that was music enough in my ears as I led Sir John and Blackett up the rickety steps of the trailer, out of the brilliant sunshine into the darkness of the cabin. Our timing was perfect. The daytime streams were at their peak, and the radiant was high in the sky so that the cathode-ray tube screen was full of echoes.

Although none of us realized it on that lovely morning, this was not to be the last occasion on which Sir John held our future in his hands. In later years my activities were to cause him much trouble and anxiety, but I know that he never for one moment regretted his initial and spontaneous enthusiasm and approval on that summer day. Over the months and years that followed this visit, both Sir John and Blackett frequently brought their distinguished visitors to the field at Jodrell, and soon they could view the cathode-ray tubes from the security of the permanent buildings that steadily took the place of the trailers scattered over that Cheshire field.

The Problem of Meteor Velocities

As our contacts with the astronomical community widened, we became aware of a major dispute concerning the origin of the sporadic meteors. Whereas there was a general consensus that the meteors manifested in the showers were moving in closed (elliptical) orbits around the sun, there was no such unified opinion about the sporadic meteors. A strong body of opinion maintained that the sporadic meteors were not permanent members of the Solar System like the shower meteors but had their origin in interstellar space. In this case they would consist of particles temporarily captured by the sun in its motion through space and moving in an open (hyperbolic) orbit around it. The issue was one of considerable astronomical importance since it bore directly on the size and mass of the particles that existed in interstellar space.

In principle the dispute could be solved by measuring the actual velocity of the meteoric particle in its motion around the sun. Calculation shows that at the distance of the earth from the sun there is a limiting velocity for a meteoric particle if it is constrained to move around the sun in a closed orbit. The limit is 42.2 kilometers per second. If a particle is moving with a velocity greater than that, its orbit is open (hyperbolic) and the particle must be a temporary visitor from interstellar space. The dispute concerned precisely this point, and we soon learned of the powerful lobby of astronomers who had interpreted their measurements as indicating that the sporadic meteors had velocities around the sun greater than this limit.

Of course, one could not go out into space to measure the velocity of the meteors; this velocity could be measured only as they plunged into the earth's atmosphere. The meteor velocities actually measured by observers on Earth were therefore compounded of the velocity of the earth around the sun and that of the meteor. The earth is moving at 29.8 kilometers per second around the sun. If the earth made a head-on encounter with a meteor particle whose velocity around the sun was at the limit of 42.2 kilometers per second, the meteor would enter the atmosphere with a velocity equal to the sum of these two separate velocities—that is, 72 kilometers per second. In fact, taking the example of this critical case of a meteor moving around the sun at this limiting velocity of 42.2 kilometers per second, we see that the actual velocity with which it enters the atmosphere can vary between the sum and difference of these two figures—that is, between 72 and 12.4 kilometers per second—depending on the direction from which it collides with the earth. Thus, one can infer the spatial orbital velocity of the meteor only from a measurement of its velocity in the atmosphere combined with a knowledge of the direction. Although the latter is relatively easy to determine, the measurement of the velocity is a different matter. The advent of the meteor is unpredictable, and the phenomenon to be observed is over in a fraction of a second.

We first learned of this serious dispute in the spring of 1947. It seemed that most of the data about the velocity of meteors had been obtained by indirect methods—such as one relying on the periodic recurrence of the meteor showers. Information about the velocity of the sporadic meteors seemed to be based on two sources. The first of these was a catalog of 611 fireballs—very bright meteors—compiled by G. von Niessl and C. Hoffmeister. More than half of these data were from British observers. Compared with that of ordinary meteors, the length of the trail of the fireball is greater, and the longer duration should facilitate more accurate measurements. This catalog, published in 1925, listed more than three-quarters of the fireballs as having velocities indicating that they were moving in hyperbolic orbits around the sun and hence must be of interstellar origin. The other source of information on the velocity of sporadic meteors was even more controversial. While at Harvard in the early 1930s, Ernst Öpik had devised an ingenious apparatus for measuring meteor velocities. This consisted of three small mirrors, two of which were made to oscillate in such a way that, when viewed through the device, the straight path of a meteor trail appeared as a looped path. The method employing this equipment became known as the rocking-mirror method, and from the shape of this trajectory, and other data, Öpik argued that it was possible to make accurate velocity measurements. Öpik used this rocking mirror in 1931–33, during the famous Harvard expedition to Arizona for the study of meteors, and in 1934 he used it at Tartu, in Estonia. In Arizona 1,436 meteor velocities were measured. In the succeeding years Öpik published a very thorough analysis of the results, from which he concluded that more than 60 percent of the sporadic meteors were moving with hyperbolic velocities and hence were of interstellar origin.

This work of Öpik's had been criticized mainly by the American meteor authority C. P. Olivier and by Prentice and J. G. Porter, the British authority on the computation of meteor and cometary orbits. Indeed, this was a classic scientific impasse since 90 percent of the 2,669 observations on which Porter's analysis rests were by thirteen British astronomers, whom he considered to be "first-line observers." Porter's conclusion was dramatically opposed to that of Öpik, and he was convinced that the sporadic meteors were confined to the Solar System, all moving in closed orbits around the sun like the shower meteors.

ASTRONOMER BY CHANCE

The Settlement of the Sporadic Meteor Dispute

I have mentioned these details to give the background to my own, unexpected engagement in the dispute about the origin of the sporadic meteors, involving nearly three years of arduous research, from September 1948 to December 1951. As so frequently happens in research where the power of new techniques is being exploited, the beginning of the affair was accidental. By the early months of 1948 several of our scientific papers had been published, and this new technique for studying meteors, unhindered by cloud, moon, or daylight, had attracted considerable attention, especially among astronomers involved in meteor studies by classical methods. We decided to invite some of the international authorities to a small conference convened in Manchester and Jodrell Bank between 7 and 9 September 1948. Of particular importance to what followed was the presence of the major contestants in the sporadic-meteor controversy—Fred L. Whipple from Harvard, Ernst Öpik, then at Armagh, and the British observer Prentice and the analyst Porter. Of course, I knew Prentice and Porter well by this time, but I encountered Öpik for the first time. He was an entirely absorbed and tenacious Estonian, conveying little effect of the years of study at Harvard or of the subsequent exile from his native land. Fred Whipple I had met in the stress of war, since he was one of the Americans who had worked in huts built perilously close to the sacred cricket field of Malvern College, the home of TRE after 1942. That brief connection led to our invitation for him to stay with us and to the beginning of a long and enduring friendship that was to prove of great importance to me in the years ahead. I had not yet lived or worked among astronomers and remember only two incidents of that visit. The first was quite inconsequential but typical of a front-rank observational astronomer. I met Whipple at Ringway airport, near Manchester—then still a small place in the country where the lights did not obscure the sky. We shook hands, and he looked at the sky. He had to orientate himself, and his instinctive means of doing so was to locate the polestar and the constellations.

The second incident was of a far more profound character. The conference was held in one of the lecture theaters of the physics department in Manchester, and a special visit had been arranged to our trailers at Jodrell. Öpik had already lectured on his rocking-mirror results, indicative of the existence of interstellar meteors. Whipple had spoken mainly about the physics of the meteor phenomenon but had also referred to his dual-station meteor photography program using cameras with rotating shutters from which he could measure the velocity and direction of a meteor and thus calculate the orbit. So far he had photographed sixty sporadic meteors, and forty-five of these produced extremely reliable data. All the orbits were closed with no sign of the interstellar component. I had been walking around Jodrell on that September morning with a small group that by chance included Whipple, Öpik, and Porter, and we had reached the vicinity of the rough corrugated iron shed we had built to protect our diesel generators. There we stopped because a violent and bitter dispute erupted between Öpik and Porter about the interpretation of the Arizona rocking-mirror results. Whipple was neutral; all that he was certain of so far was that his photographic technique did not reveal the existence of interstellar meteors—but they were bright meteors, and Öpik's conclusions were based on a sample of much fainter meteors. Whipple was not prepared to take sides without adequate evidence on the interpretation of the results of a Harvard expedition. He did, however, turn to me and, pointing to the trailers, remarked that we ought to be able to settle the question by using the new radio method for measuring the meteor velocities. So far this equipment had been tested during the major meteor showers, where the velocities of the meteors were already known to within small limits. This had demonstrated clearly that the radio measurements were sound, and now Whipple was suggesting that we turn our attention to the sporadic meteors. My response was that we could not do this, because we had no means of determining the direction of flight of the individual meteors whose velocities we measured, and that this would be essential if we were to find out whether the orbits in which they were moving were open (interstellar) or closed (Solar System). Whipple's response was that we did not necessarily need this amount of detail. He said that we should measure a large number of sporadic meteor velocities, and if there was a hyperbolic (interstellar) component, a certain number of these must have velocities greater than 72 kilometers per second. If we did not find any velocities greater than 72 kilometers per second, it followed that there could be no interstellar meteors. I asked Porter and Öpik if they agreed that this would provide a definitive settlement of the problem, and upon their concurrence I remarked, with an optimism that was to be entirely misplaced, that the problem would be solved by Christmas. Our ability to settle this question depended on a new method for the accurate measurement of meteor velocities, a method we had recently developed. The idea arose in a discussion with Nicolai Herlofson. Although not a member of my small group at Jodrell, he was still working in Blackett's department at that time and had made a thorough investigation of the physical processes by which the small meteoric body produced the long, thin column of light and ionization as it plunged into the atmosphere.

Herlofson knew of our theoretical work on the process by which radio waves were reflected from the meteor trail, and he had made the ingenious suggestion that we should develop a method by which the variations in the intensity of the received signal could be measured as the ionized trail was actually being formed. This intriguing idea had a simple optical analogy related to the rhythmic variation in brightness at the edge of a shadow cast on a screen when an obstacle is placed in the path of a beam of light. The explanation of this diffraction effect was first given early in the nineteenth century by the French physicist Fresnel. In the case of the meteor trail the range could be measured accurately by the radar equipment, and thus the width of the zones could be calculated. If we could then measure the rate at which the maxima and minima occurred in the reflected radio wave, we could measure precisely the actual velocity of the meteor. If we were to measure velocities of tens of kilometers per second by this method, we had to measure the amplitude of the radio wave reflected from the trail every thousandth of a second. We soon realized that this information was buried in the echoes we were already observing. Our transmitter was a typical radar transmitter. It transmitted not continuously but in short pulses. We were radiating six hundred of these pulses every second. The signal we received after it was scattered from the meteor trail was similarly pulsed at this rate, but our cathode-ray tube display simply piled all of the pulses on top of one another to give a single echo. If we could devise a means of spreading out these pulses and photographing them on the cathode-ray tube so that the strength of each returned pulse could be measured, it seemed that we would have a revolutionary method for the accurate measurement of meteor velocities.

The adjustments to "spread out" the echo on the cathode-ray tube were simple; the real problem was that there was no warning of when a meteor echo would appear. The only hope was to make the echo itself initiate the process of speeding up the time base to take its own pulse-by-pulse photograph. The problem was solved by two young students who had recently joined my Jodrell group to pursue their postgraduate studies. The first to arrive, in December 1946, was a young New Zealand scientist, C. D. Ellyett, who had been given a National Research Scholarship and leave of absence from the physics department of Canterbury University College to study in England during the academic year 1947–48. By the time he returned to New Zealand, in November 1949, he had been a partner in one of our significant developments and had gained the experience that enabled him to initiate a major series of meteor observations in the Southern Hemisphere. After he left Jodrell, I did not see him again for thirty-five years. It was in Sydney, but the cold day in December 1946 when I first greeted him in the field at Jodrell and the exciting years that followed remained vivid memories. In January 1947, only a few weeks after Ellyett arrived, another young man joined my group. The name of J. G. Davies will appear frequently in the following pages. He had graduated in 1944 and had been sent to work on proximity fuses at the Royal Aircraft Establishment. Entirely accidental circumstances through a slender connecting link with a member of my wartime group led to his arrival at Jodrell.

These two young men soon devised and built the electronic equipment so that the first echo pulse received from a meteor would trigger a fast sweep of the time base of a cathode-ray tube. A camera with an open shutter was mounted to photograph the tube, and the sequence of operations was arranged so that the camera film was wound on automatically for the next event. A serious problem arose as soon as an attempt was made to use the device operationally. The electronic trigger had to be extremely sensitive, and unfortunately the device was triggered by random noise impulses that had nothing to do with the meteors. This was a severe hindrance, and several months elapsed before Davies and Ellyett overcame it by devising an electronic discriminator that could distinguish between the wanted impulse from the meteor trail and the unwanted random noise impulses.

By coincidence this apparatus was ready for testing precisely two years after I had first switched on the equipment in the trailers at Jodrell. It was the night of 13–14 December 1947, the peak of the Geminid meteor shower. At 3 A.M. a dripping film rushed out of an improvised darkroom revealed the beautiful pattern of the Fresnel zones formed by the ionized trail of a Geminid meteor as it plunged into the atmosphere. Of course, by today's standards all of this would be regarded as elementary, but by the techniques of that immediate postwar period this was a brilliant accomplishment by Davies and Ellyett, carried through in the unsophisticated environment of the trailers in a country field. Although the discrimination worked well enough to enable meteor velocities to be measured, the difficulty of such work under these conditions was considerable. For example, in one sequence during the autumn of 1948 only 1,700 frames were triggered by meteors, out of a total of 14,500.

Nine months after the successful test of this velocity-measuring system on the Geminid meteor shower, Whipple suggested we should use it to settle the controversy about the interstellar meteors. I asked J. G. Davies to work on the problem with me, and we soon enlisted the aid of another young postgraduate student, Mary Almond. We thought that the experiment would be quick and simple. We would direct the searchlight aerial toward the east for a few hours around six each morning so that the meteors we recorded would be those making a nearly head-on encounter with the earth and thus give the greatest chance of recording velocities in excess of the 72-kilometers-per-second limit, if they existed. On 18 September, less than two weeks after the confrontation of

Öpik, Porter, and Whipple, we began work at five and continued every morning until 18 December, by which time we had accumulated 1,444 camera frames, showing echoes in 230 hours of observing. From these we were able to measure the velocities of sixty-seven meteors. The distribution of the velocities agreed well with that to be expected theoretically, on the assumption that there were no interstellar meteors. However, we were not entirely happy, since it seemed possible that our method might be biased against recording meteors of high velocity. Therefore we decided to repeat the measurements with apparatus working on a longer wavelength. It was October 1949 before this new equipment was complete, and from 10 October to 20 December we once more began recording every morning at five. The longer wavelength gave us much more sensitivity, and in 42 hours of operation we recorded 3,045 echoes, from which 187 velocities were measured. Again the distribution showed excellent agreement with the theoretical curve assuming that no interstellar meteors existed.

Now we were satisfied, and I wrote to the major partners in dispute to convey the news of our results. Naturally I expected an enthusiastic response. This certainly came from Prentice and Porter, who were pleased that their own conclusions about the absence of interstellar meteors was confirmed. Whipple, too, accepted our conclusions but wondered whether the same result would be obtained from the fainter meteors, which were more likely to have an interstellar origin. On the other hand Öpik rejected our results absolutely and violently, on the grounds that we were still biased against recording the meteors of high velocity, which he had measured with his rocking-mirror device. Hoffmeister, too, when we finally managed to correspond with him indirectly in East Germany, objected to our results on the grounds that they were biased against the recording of high-velocity meteors.

We had already satisfied ourselves that our measurements were not selective, but in order to demolish the Öpik-Hoffmeister arguments, we began another series, this time in the spring months with our aerial beam directed to record the sporadic meteors that would have the lowest, instead of the highest, velocities when they entered the atmosphere. This was a somewhat more pleasant exercise since it entailed working in the afternoon and early evenings of the spring of 1950. From 18 February to 29 April we measured 87 velocities from 1,706 meteor recordings. Once more our results were conclusive—there were no interstellar meteors.

Finally we had to overcome Whipple's worry that the fainter meteors—that is, those below the visual and photographic limits—might be interstellar. To do this we built a larger aerial and increased the power of the transmitter. In the early autumn mornings of 1950 we measured 335 velocities, from 2,824 recordings of echoes, and during the spring evenings of 1951 another 57 velocities, from 1,586 echo frames. Again our results were conclusively against any interstellar meteors. We had now covered all the magnitude ranges appropriate to the visual and photographic measurements and had extended the measurements to meteors that were several magnitudes fainter than the visual or rocking-mirror limits.

Finally, in the early dawn hours of November and December 1951, we still further increased our transmitter power and overall sensitivity and measured another 362 velocities from 2,181 echo frames. We had gone three or four magnitudes fainter than the faintest meteors visible by the unaided eye, well into the "telescopic" meteor range of magnitudes, and again our results were decisive.

This investigation, which at the time of the dispute in September 1948 I had firmly believed we could complete by that Christmas, had extended into the third Christmas. We had involved ourselves in processing several thousand feet of 35-millimeter film exposed in over 575 hours of observation, mostly under the unpleasant conditions of the predawn autumn hours, and had to analyze 12,786 film frames of meteor echoes to obtain 1,095 velocities. Our detailed results and analysis were published in a series of papers by the Royal Astronomical Society over the years 1951–53, by which time other workers, in Canada, had confirmed our conclusions and Whipple had begun his own, massive accumulation of photographic and radio echo meteor data at Harvard, none of which showed any sign of interstellar meteors. In the conclusion of our final paper we wrote that we had initiated the work in an attempt to settle the controversy over the origin of the sporadic meteor component and that the result was entirely in favor of the view that the sporadic meteors are permanent members of the Solar System.

Any reader of this story of the interstellar meteor problem is bound to ask why the results of Hoffmeister and of Öpik turned out to be so erroneous. It is a valid query, one by no means unique to this particular sequence of events. Indeed, there is rarely a simple answer, and the issue under discussion here is no exception. Hoffmeister's argument depended on two separate investigations. First was the fireball catalog, compiled jointly with Niessl. As I already mentioned, the false conclusion that more than half of the fireballs cataloged must have had an interstellar origin undoubtedly resulted from the use of data far less accurate than the observers of the fireballs believed. In our own correspondence with Hoffmeister, during the course of the Jodrell Bank measurements, a different issue arose. Apart from his association with the fireball catalog, he had by then reached a similar conclusion about the interstellar origin of sporadic meteors from quite different considerations. This conclusion was based on a study of the diurnal and annual variation of the number of sporadic meteors seen by visual observers. Hoffmeister worked in East Germany, and at that time direct contact with him was impossible. Fortunately, with the assistance of an intermediary, we found a route by which we could exchange correspondence and data with him. Eventually we managed to place in his hands our unpublished systematic radio data on the diurnal and annual variation of the numbers of sporadic meteors. When he considered these results, he realized that his analysis had been based on visual data that were not inaccurate, like the fireball data, but inadequate (particularly since it necessarily omitted any details about the daytime activity). When he applied his analysis to our records, the evidence for interstellar meteors disappeared, and he graciously acknowledged this fact.

The case of Öpik and his rocking-mirror results was quite different. We had ample opportunity for discussions with Öpik, but he would never acknowledge that his results were erroneous, and no single reason for the discrepancies was ever agreed between us. The contrast with Öpik's results could scarcely have been greater. At that period I made a detailed study of his work and of the various criticisms made by other workers. Öpik was an outstanding theorist, as the bulk of his published work subsequently revealed. In my experience it is rare for a scientist to be both a first-class theorist and a front-rank observer, and he made the cardinal error of omitting to prove the accuracy of his device on a well-known meteor shower. In the absence of such a fundamental check the validity of the basic observational data had to be accepted. To these, Öpik applied conventional, but highly complex, stellar statistics. As far as I am aware, no one has ever attempted a detailed investigation of how the gross errors emerged from this treatment. Nevertheless, it is not the only example in science where the application of valid statistical methods to poor observational data has yielded false results.

This episode of the sporadic-meteor component, which originated casually during the visit of Whipple, Öpik, and Porter to Jodrell on that September morning in 1948, proved a lengthy and illuminating experience for those of us who were newcomers to the domain of astronomy. I had often been engaged in bitter arguments with colleagues during the war years on operational issues where matters of life and death were involved, but I had not realized that a similar bitterness could develop about beliefs on the nature of the interstellar domain. For me it was a useful experience. Many times in the future I would encounter these violent divisions and also discover that in some cases their origin had a significant emotional basis. Sets of scientific data may often seem complete and decisive but prove to be open to various interpretations. If I did have any personal emotions about the sporadic-meteor problem, they were in the nature of regrets. It is never very pleasant to prove that a scientist is wrong, and I did not enjoy this involvement with Öpik, a refugee from his native country. Neither was I particularly thrilled with the nature of our conclusions, since I always felt that it would have been far more interesting and important if the sporadic meteors had been of interstellar origin. However, by the time we had concluded our work on this issue, I was deeply involved in matters that were to have a dramatic impact on our future and rapidly lead the whole of our researches away from the problem of meteors.

9

The First of
the Large Radio Telescopes

The description of our work on the sporadic meteors may give the impression that, although it was carried out under arduous conditions, the pervasive environment was calm and academic. In fact, it was turbulent in the extreme, and in the ensuing dramas my own career and the very existence of Jodrell narrowly escaped destruction. I have mentioned that in the late summer of 1945, when Blackett returned to Manchester, he instantly wrenched me away from my attempts to restore the prewar cloud chamber equipment to working order and reminded me of our wartime paper and the promise that we would use the radar equipment to develop a new method for studying large cosmic-ray air showers. At that time we both realized that our assumptions and calculations had been of an elementary nature. Before there could be any question of my begging radar apparatus from Hey or anyone else, Blackett insisted that I make a thorough study of the problem now that libraries were at hand and I had time to do so. The first of these studies is headed "July 1945. PMSB [i.e., Blackett] suggests (a) theory of shower formation (b) diffraction theory (c) atmospheric scattering theory, effect of collision damping etc.—Problem is—what happens when a pulse of e.m. [electromagnetic] energy, wavelength λ goes into the atmosphere and in particular when it strikes ionization due to a cosmic ray shower?'' There follow fifty pages of formulae and calculations, and the final numerical estimates show that, with the equipment I proposed to borrow, the ionization from large cosmic-ray showers should produce a radar echo. With this confirmation of our more elementary wartime calculations, I borrowed the ex-army trailers and first operated them in the quadrangle of the university on 12 September 1945 and subsequently moved to Jodrell Bank in December, as I have recounted.

In retrospect I think it may have been fortunate for our future that after the initial tests at Jodrell Bank, in mid-December 1945, I was absent for several weeks with a minor illness and did not return for consistent work with the apparatus until the end of January 1946. It was fortunate because this shifted the real crisis of my affairs with Blackett into the early summer and much closer to our first successes with the meteor observations of the Perseids in August and the Giacobinids in October than would otherwise have been the case.

I cannot place with precision the time at which this crisis began to develop, because a key letter sent to Blackett has been lost. Even so, its contents are ingrained in my memory. It was from T. L. Eckersley, a distinguished British ionospheric scientist. Eckersley had remembered that paper, published by the Royal Society early in the war, in which we had suggested that some of the short-lived radar echoes could be scattered from the ionization caused by large cosmic-ray showers. We had made a small mistake, Eckersley kindly pointed out. There was a factor of $\frac{8\pi}{3}$ that we had omitted from a formula well known to every student of physics. Blackett was exceedingly annoyed that I had involved him in this mistake.

Nearly thirty years later I had the task of writing the Royal Society biographical memoir on Blackett, and only then did I realize why Blackett was so agitated at that time. It was at the moment of crisis in his epic dispute with Prime Minister Clement Attlee and the small committee deliberating on Britain's atomic program and its attitude to international control. Blackett had been alarmed about the policy in the United States and the talk there of fighting a preventative war before the Soviet Union developed the atomic bomb (the first Soviet atomic bomb was not tested until August 1949). He had circulated a minority paper to the inner circle of the Cabinet urging a neutralist policy for the United Kingdom and had been rebuffed by Attlee, who wrote across his memorandum, "The author, a distinguished scientist, speaks on political and military problems on which he is a layman."* Attlee had sent this to the British Chiefs of Staff. Although I could know nothing of that at the time in 1946 to which I refer, Blackett was being forced into isolation from the advisory machinery of the government, and he was very angry.

In the event, I could not understand his deep annoyance over the factor of $\frac{8\pi}{3}$ omitted from the draft of a paper I sent to him, written on an airfield under the stress of war and without access to a library. That factor could make no significant difference in the task of detecting echoes from the cosmic-ray showers. He gave me Eckersley's letter and suggested I write to him and explain how I had inadvertently associated Blackett with this error.

Although Blackett apparently had not realized this, the real force of Eckersley's letter was not the minor business of the $\frac{8\pi}{3}$ but the last sentence, "By the way would you ask Lovell if he has considered the effect of the damping factor." The day I had posted the draft of that paper to Blackett we were being machine-gunned by German aircraft on our cliff site at the establishment in Dorset, and it was no time to consider effects of damping factors high in the atmosphere. Neither was I aware of the catastrophic effect of this factor until Eckersley's query made me look into the details. In principle the issue is quite a simple one. The scattering formula I had used for the calculations assumed that the electrons were completely free to oscillate under the influence of the incident radio wave. If they were hindered by colliding with the atoms or molecules of the air in the vicinity, they could not execute their full motion, and the effect would be to reduce the amount of energy they scattered back to the earth. Consequently it would become far more difficult to detect the radar echo.

I could not have been entirely unaware that such a collision-

---

*Margaret Gowing, *Independence and Deterrence. Britain and Atomic Energy 1945–1952*, vol. 1, *Policy Making* (London: Macmillan, 1974), pp. 172, 194–200.

damping effect occurred, because in my July 1945 calculations the final section is headed "Neglecting Collision Loss." However, the reading of Eckersley's letter to Blackett sent me back to the library, and I soon developed a great anxiety. The assumption I had made that the electrons would be free to oscillate was very far from the truth. In fact, the frequency of the collisions of the electrons with neutral molecules at the altitudes at which I expected to detect the cosmic-ray showers would be very high compared with the frequency of the radar transmitter I was proposing to use. At sea level, under the conditions I had assumed, there would be a trillion collisions per second, and at an altitude of twenty kilometers there would be a hundred billion collisions per second. The effect on the proposed experiment would be catastrophic. I frantically covered endless sheets of paper with calculations refining all the many elementary assumptions about the nature of the cosmic-ray shower ionization and searching for any change of the radar characteristics (such as the wavelength) that would make the research feasible. Recently I discovered the long paper that I had finished by the end of January 1946 but that had never been published; its first conclusion was "If these calculations and assumptions are correct the experiment is unlikely to succeed."

Blackett was never one to suffer fools gladly, and when I showed him these calculations he left me in no doubt of what he thought of my scientific ability. Fortunately for me he was heavily involved with his own scientific problems at that moment in addition to his political confrontations with the prime minister in London. This episode now appears as a critical test of my relationship with Blackett; if he had then had the time to concentrate on my inadequacies, I doubt that our association would have survived. In the event, his own opinions on atomic matters were being disputed at the highest levels of government, and the next time we met he was full of the inadequacies of his contemporaries on the Cabinet advisory committee and of his own despair at the U.S. policy. Many weeks elapsed before he once more concentrated on my problems, and by that time I had finished even more calculations. In fact, I had pointed out to Blackett that my conclusions also contained a rider to the effect that I had based the calculations on various assumptions concerning the spread of the cosmic-ray showers in the atmosphere—that is, the volume of the electrons that would be effective in scattering the radio waves back to earth. Fortunately the Hungarian L. Jánossy, with whom I had worked before the war, was still in Blackett's laboratory, and he was an authority on this matter. After many discussions with Jánossy I wrote yet another note, headed "New Calculation of radio echoes from showers," and gave it to Blackett. In this note, acting on Jánossy's advice, I had simplified the previous calculations by neglecting the spread of the shower particles and by assuming that, for the wavelength I proposed to use, the ionization produced by the shower could be treated as a "point cluster"—that is, by assuming that 4 meters, the wavelength of my transmitter, would encompass the major core of the shower particles. This made a huge difference in the previous calculations, and it seemed that if we could improve the sensitivity of our equipment by several thousand times we might observe one echo from a large cosmic-ray shower every twenty-four hours on the average.

It was May 1946 before I had finished this note, and when Blackett saw it he insisted that I send him a detailed report of everything that had happened since the trailers first appeared in the university, in September 1945.

In fact, I was soon to move away from this crisis, which nearly obliterated any scientific development at Jodrell Bank. In chapter 7 I referred to J. A. Clegg and C. J. Banwell, who bore much of the credit and responsibility for helping me out of the crisis of those early months of 1946. It was indeed my great good fortune that at that critical moment I had two such companions to join me in the trailers—one an expert on aerials and the other on electronics—with whom I could discuss how to improve the sensitivity of the equipment. Faced with this situation, we found only three possibilities for obtaining the required increase in sensitivity of thousands of times. One was to increase the transmitter power. But that we could not do, because we had no money to purchase a superior transmitter. My idea that we should beg one of the extremely powerful coastal defense radar transmitters came to nothing, because they were still in operational use, and the operation of one of these at Jodrell would have required an army of technicians. In any case, we concluded that this would be an inefficient procedure since the improvement would appear only as the square root of the transmitter power. This left the receiver and the aerial. As regards the receiver, Banwell said he could make only a small improvement because the sensitivity would be limited by the cosmic noise. Although I did not realize it at the time, it was a comment laden with significance for my future and that of Jodrell. There remained only the aerial system, and without delay Clegg drew up the plans for an enormous broadside array that would have a power gain of more than five thousand times over our existing aerial.

The fifteen-page report that I wrote for Blackett about the work to 21 June 1946 was a detailed account of all the echoes I had observed since 12 September in Manchester and since 14 December at Jodrell Bank. By an analysis and argument that now seems tortuous I had persuaded myself that, in the case of a few special echoes lasting for about ten seconds, it was "not out of the question that this type of echo may be caused by very high energy cosmic ray particles." Perhaps this persuaded Blackett to let us continue with the plans I outlined for improving the equipment. The critical component was the aerial, and under that heading, although I did not suspect it then, I was determining the future of Jodrell Bank. I referred to three aerials. The "searchlight aerial," which was then almost complete and which, as already described in chapter 7, we were to use with such success during the Giacobinid meteor shower in October of that year. The second reference was to the "big paraboloid of diameter 75 yards," which we were about to construct and which as far as was known would be "the biggest paraboloid ever constructed." The third was a proposal for a huge "broadside array," part of which we had already built.

The Aerials

The searchlight aerial was completed in time for the observation of the Giacobinid meteors in October 1946, and the significance of its use on that and later occasions will have been evident from the accounts in the preceding chapters. The broadside array was a different matter altogether. A photograph taken in midsummer, at the time of my report to Blackett, shows the half-completed searchlight aerial and, nearby, a tall tower constructed of scaffolding tubes.

When I reflect on that photograph, I am astonished that in collaboration with Clegg alone I started on that considerable engineering construction with no professional competence to do so. It was our aim to provide a framework a hundred feet high and of about the same width on which to build the array of dipoles that would form the aerial. There was a strict rationing of timber and steel; in any case, we had no money to buy those materials. At that time, however, the bomb damage was being repaired and new construction was beginning everywhere. A journey to almost any city would reveal large quantities of steel scaffolding tubes, and we discovered that we could lease these tubes quite cheaply. A large quantity of them was soon deposited in our field. The fixing arrangements to join one tube to another or to make a framework was quite simple, but we soon discovered that if our tower was to be stable we would have to fix it to the ground in concrete. We had no mechanical mixer; neither did we have the sand, cement, and stones. On obtaining these ingredients, we started the excavation and mixing for eight solid concrete blocks by using spades. In this amateurish manner we eventually produced a part of the intended framework, but it was never completed. In fact, shortly after my report to Blackett in June the entire idea was abandoned. We thought of a much better use for the scaffolding tubes—we would use them to make the framework for a giant dish-type reflector in the shape of a paraboloid. This would need only one dipole antenna system at the focus. The changing of the wavelength of operation would therefore be far simpler in this case than for the broadside array, where large numbers of dipoles and their connections would have to be changed if we needed to alter the wavelength. During the summer of 1946 we worked out our plans. That is to say we decided to make the reflecting bowl from a grid of thin wire supported on a system of steel cables held in place by the scaffolding tubes.

155

There were never any plans in the conventional sense of a series of drawings with the various strengths and tensions calculated— we simply used our intuition. Our guiding principle was that the aerial must be as large as possible since we wanted the maximum sensitivity for the cosmic-ray experiment. That was the easiest of the problems to settle because in the middle of the field the Park Royal vehicle was immovably sunk into the mud, and the distance between that vehicle and the boundary hedge of the field was about eighty yards.

In effect we were planning to build a giant wire saucer, using scaffolding posts around the perimeter on which to anchor the reflecting wires. The next problem was that we must be able to reach the tops of these posts in order to do the fixing, and our only means of getting there would be ladders. After some trials as to the height at which we felt able to work on a ladder, we fixed it at 24 feet. We intended to fix these perimeter scaffolding tubes into concrete blocks, but they would need tying by steel cables to anchor blocks outside the perimeter in order to balance the pull toward the center of the main wires forming the framework. This, in turn, determined the eventual size of the wire bowl—it could be 218 feet in diameter.

Now, a paraboloid, whether a searchlight mirror or a radio device, has to be fed from the focus—by the light source in the former case or by a dipole or small aerial array in our case. The distance of this focal point from the apex of the paraboloid is determined by the shape of the paraboloid. In our case the depth of the paraboloid (24 feet) was going to be small compared with the diameter (218 feet). The bowl would be a shallow one, its focus to be 126 feet above the ground. In order to make it possible for us to construct the reflecting wire bowl, we had given ourselves the difficult problem of fixing the feed dipole 126 feet above the ground at the center of the paraboloid. We were halted in our plans until Clegg remembered that the wooden towers holding the receiving aerials of the coastal defense radars were about this height. One more appeal to our wartime colleagues at the Air Ministry, and Clegg was soon driving a large truckload of wooden struts into Jodrell. We constructed a concrete base at the center point of our proposed paraboloid and proceeded to erect the tower. By now the autumn of 1946 was advanced, and with the tower built to about 20 feet our enthusiasm began to wane as the shortening days made the task of reaching the 126-foot level seem even more formidable and dangerous for two inexperienced people. Term had begun, and with many lectures to give we needed some good fortune—and it came in a most unexpected manner.

Blackett was in close contact with the Department of Scientific and Industrial Research (DSIR), through which the government made grants for university research projects of "timeliness and promise." One Friday—a day on which I always had a full teaching commitment in the university—Blackett introduced me to C. A. Spencer, the member of the DSIR in charge of research grants to the universities. Blackett, having finished his business with Spencer, asked me to take charge on the Saturday morning and suggested that he might be interested in visiting Jodrell on his way to the London train. On that morning I showed Spencer some meteor echoes in our trailer. It happened by chance that anyone descending the steps from the trailer would be facing the base of the wooden tower that Clegg and I had constructed. Inevitably he inquired why we were reconstructing this former radar tower in the bare field. I explained our plans for the paraboloid and without any thought of success said that we did not yet have enough money to buy all the material we needed; in particular we required a thin, 126-foot steel tube to support the aerial at the focus, since we doubted whether we were capable of finishing the wooden tower to that height. "But this is exactly the kind of research of timeliness and promise that the department likes to support," Spencer remarked. Thus I unexpectedly applied for £1,000 to complete the building of the paraboloid and got it without hesitation. In that pleasant way I was introduced to the grant-giving mechanism of the DSIR and little realized on that Saturday morning how traumatic our relationship would become in a few years' time.

The first charge on this money was to buy a tubular steel mast that would support the aerial at the focus 126 feet above the ground. We consulted the radar aerial experts in TRE, who advised us that the engineering firm of Coubro & Scrutton, in

London, might be able to help us. I was soon in touch with F. D. Roberts of that firm, and he satisfied our needs exactly—a thin, 126-foot tube, pivoted on a concrete base at ground level and secured by guy wires. With this problem solved we now faced the task of constructing the paraboloidal bowl. The first operation was to fix the twenty-four perimeter posts of scaffolding tubes. From each of these posts we planned to stretch radial runs of heavy steel cables to ground level at the center. However, this would not have been a shape close enough to the paraboloid we needed, so we also planned to tie each of these cables to further concrete blocks at equal intervals of 27 feet 4 inches from the center. We calculated that by this means we could make a framework that would nowhere deviate from the true paraboloid by more than 5 inches, and that accuracy would be satisfactory for the wavelength of 4 meters that we proposed to use. The tie points along the twenty-four radial cables required seventy-two concrete blocks to be sunk into the ground, the perimeter posts required another twenty-four, and a further forty-eight anchor posts were deemed advisable to secure the rigidity of these outer tubes. For two people with spades it seemed a formidable task, but fortunately the resolution of Clegg and myself was not to be tested quite so severely.

When we were ready to start again on this constructional task in the early spring of 1947, one of my university colleagues mentioned that someone had been appointed to a university workshop as a technician but that he might be more suited to the engineering tasks we were undertaking at Jodrell. In that way I acquired Edward ("Ted") Taylor. The presence of a skilled worker like Taylor at Jodrell Bank in those days was a major piece of good fortune. It was not that skilled workers did not exist in the country—the simple fact was that all skilled people were either still held in their wartime posts or were being called up for essential work as though the war were still being waged. Indeed, it is all too easy to forget the problems that beset us in those immediate postwar years.

It was against this national background that I was so fortunate at this critical stage in 1947 to be given Ted Taylor to help me at Jodrell. In the event, I soon realized that under any circumstance any man who got Taylor to help him would be fortunate. In the years before the war, having served his engineering apprenticeship, Taylor was put by his father on a ship leaving Liverpool for New York, carrying only his bag of tools. He made good in Philadelphia, returned to England shortly before the outbreak of war, and was soon the foreman overseeing the construction of Rolls-Royce aircraft engines at a temporary factory in Trafford Park, Manchester. The need for this factory ended with the cessation of hostilities, and so Taylor found his way to the university and to Jodrell. In the years to come Taylor created and took charge of our own workshop and remained by my side until he retired decades later. In those early spring months of 1947, though, he faced the problem of the scaffolding tubes and the concrete blocks. He had all the right contacts, and soon a concrete mixer was adding to the noise of our diesel generators, and the face of that field was rapidly transformed.

I should perhaps mention at this point the continual help we received from Moston, the farmer who had first managed to start the diesels in December 1945. Throughout these early years he would as often be found working with us as doing his farm work, and we soon had an illuminating illustration of why he felt able to do this. One of our radar transmitters was sited at the boundary hedge, its aerials pointing over a field in which he grew grain. That year he reaped a tremendous crop and maintained to all that this was owing to the beneficial influence of our transmissions. The next season we agreed to make a more scientific test. By that time I had consulted the botanists in the university, who informed me that little was known about the effects of such radiation. They agreed to carry out appropriate statistical tests on the yields of grain from the radiated and nonradiated fields. The results were puzzling because the yield from the radiated field was now far less than that from the other fields. Then it transpired that Moston had undermined all such scientific tests by deciding that our radiation treatment was more effective than his fertilizers and so had ceased using them.

It was this affair that predisposed Moston to be so generous of his time and effort in helping us, and Taylor was quick to take advantage of his help and of his local knowledge. The business of the wheat radiation also led to another interesting experiment in 1947. In the course of normal family contacts, I occasionally met F. Gordon Spear, my wife's uncle, a distinguished senior staff member of the Strangeways Research Laboratory, in Cambridge, and an authority on living cells. He was naturally interested in my developments at Jodrell, and when I casually told him about the radiation of the wheat field he immediately asked if he could carry out some controlled tests on his own cell material. In late November of that year he exposed several dozen preparations in our beam. He said that some work had recently been done elsewhere but that the results had been confused by the heating effects of the radiation on the tissue, and he was careful to avoid this effect in the tests. In early December he sent to me a note about the results—the irradiated cells had shown a slight and "possibly significant" increase in cell division compared with the controls. Gordon Spear had been anxious to use radiation in the 10-centimeter region, but at that time we had only the 4-meter transmitter. Unfortunately circumstances developed such that this work could not be resumed when we eventually had transmitters in the centimeter-wavelength region. Perhaps it was as well for the sake of Jodrell that we were not led further into this biological maze, because I believe that even today there is serious dispute about the possible effects of such nonionizing radiation on living cells.

By the time Gordon Spear carried out this experiment in November 1947, the 218-foot aerial had been completed. The twenty-four perimeter posts had been secured, the ⅜-inch-diameter steel cables from the top of each of these posts had been fixed to the ground at the center near the foot of the 126-foot tubular steel mast, and the radial runs had been strained to the ground at three points so that each of these points lay on the paraboloidal surface. On this rigid but open framework of steel cables we now faced the problem of forming the reflecting surface of the paraboloid.

Ideally the reflecting surface of a paraboloid for use in the radio wave band should be a continuous sheet of conducting metal. Radio waves of any wavelength would then be reflected from the surface without any loss through leakage. If the surface is not solid but is made of open mesh, for example, there will be loss of efficiency because radio waves will leak through the mesh. The amount of leakage depends on the wavelength and the mesh size. We had originally planned to fix wires on the supporting framework to make a 4-inch mesh surface. The mesh size would then have been small compared with our wavelength of 4 meters, and the loss of efficiency through leakage would have been negligible. In the event, in 1947 we left this as an intention for the future and decided to test the device by fixing wires at 8-inch spacing—not in the form of a mesh. By using the appropriate polarization (that is, with the dipoles at the focus mounted parallel to the run of these wires), we would limit our leakage through the surface to only 5 percent of the power on this wavelength of 4 meters. We therefore proceeded to purchase eight miles of this 16-gauge galvanized wire and faced the task of winding it onto the web of cables. This was simple enough for the section of the bowl near ground level at the center; as we moved out to the perimeter, though, the height of the steel cables above the ground necessitated the use of ladders, and this made the work very slow and laborious. We then discovered that the top of one of the covered trucks we had acquired was firm enough to stand on. This solved the problem of laying the wire in the perimeter region since one of us would drive the truck slowly under the framework while a person on top would loop the wire over the cables. This wire now had to be secured to the main steel cable framework, and that meant about five thousand secure ties. We were desperately anxious to finish the job before the short and cold days of the winter arrived, so at weekends we pressed wives and children into the task of making these secure ties that could be reached from the ground or from boxes or stepladders, while we concentrated on the high tie points near the perimeter.

We now had to mount the feed aerial at the focus, and that was at the top of the mast. We had asked Roberts of Coubro & Scrutton how this access could be arranged, and, as a person well acquainted with tall masts, he casually said that it was the usual practice to ascend in a boatswain's chair. Steeplejacks and sailors would no doubt have taken this operation in their stride, but for us it was a rather terrifying prospect. However, we had built the small array of dipoles and mounted them on a 5-inch-diameter wooden pole that could be pushed into the top of this tubular mast.

In September, Victor Hughes, who had just graduated in Manchester, joined me to study for a higher degree, and with his help in the darkness of a damp November evening we completed the connections and switched on the transmitter. This paraboloid was a great success—but not for the primary purpose for which we had built it. We had made a simple, common transmit-receive device that enabled any such aerial to be used simultaneously for both transmission and reception, and we had hoped that the power gain of this huge aerial would reveal the occasional echoes from the cosmic-ray showers. This was still the primary reason for our presence at Jodrell Bank. We spent many hours in front of the cathode-ray tube display but, apart from occasional echoes from meteor trails, saw no evidence of any new phenomena.

That primary intention was never fulfilled, but the secondary purpose quickly dominated the use of the "transit telescope," as it became known. Even so we had no time for regrets or disappointment, because we soon recognized that we had built a major instrument for the study of radio waves from the universe. Indeed, this, the first of the large radio telescopes, was to shape our future destiny. It was a beautiful and elegant instrument, the slim mast emerging from the Cheshire countryside as though it were a great finger directing our thoughts and efforts to the remote regions of time and space.

10

Radio Astronomy
—A New Science

In the previous chapter I described the situation in the summer of 1946 when it was realized that a large improvement in the sensitivity of our equipment would be needed if there was to be any hope of proceeding with the cosmic-ray experiment. The easiest solution seemed to be to improve the receiver, but Banwell said this would make only a small difference because the limit would be set by cosmic noise. This phenomenon was soon to be the major concern of the research at Jodrell and, indeed, the primary reason for our continued development. It is therefore curious that Banwell's remark was the first I had heard of this subject.

In retrospect this seems a remarkable confession of ignorance because the existence of the cosmic noise had been discovered accidentally by Karl Jansky in the early 1930s. Jansky was studying atmospherics at the Bell Laboratories in New Jersey. He was investigating the atmospheric disturbance to long-distance radio communication and found that there was in his receiver a continuous background noise, which varied in strength throughout the day. The noise reached maximum strength once in the twenty-four hours, but with brilliant insight Jansky recognized that this maximum occurred every twenty-three hours and fifty-six minutes. This is the period of the earth's rotation with respect to the stars, and Jansky was aware that the source of this noise, or "cosmic static," must lie in regions of space outside the Solar System. Jansky did not pursue this discovery in detail, but Grote Reber built a receiving system as a spare-time occupation in his yard at Wheaton, Illinois, and confirmed Jansky's conclusion. Reber built a small-diameter paraboloidal dish for his aerial system—an arrangement that in later years became familiar as the conventional form of radio telescope. He investigated the strength of the radio emission from different parts of the sky and concluded that it originated in the Milky Way in the rarefied regions between the stars.

Jansky published the accounts of his discovery in an electrical-engineering journal, and the first of Reber's accounts in the *Astrophysical Journal* did not appear until Europe was ravaged with war. Even those scientists who knew of this work did not consider it of great importance to the study of astronomy. Hitherto it had been possible to study the universe from earth because the light from the stars and distant nebulae was not significantly absorbed or scattered in the earth's atmosphere. By an accident of evolution our eyes are sensitive to just that small region of the electromagnetic spectrum, extending from the short-wavelength blue to the longer-wavelength red, which can penetrate the obscuring layers of the atmosphere. Since it was known from the laws of physics that hot bodies such as the sun and stars, at a temperature of thousands of degrees, emitted the peak of their energy (known as thermal radiation) within this wavelength region, there was no reason to suspect that the obscuration of the atmosphere outside the limits of the optical spectrum imposed any serious limitation on our knowledge of the universe.

The remarkable discoveries made by Harlow Shapley and Edwin Hubble with the large optical telescopes on Mt. Wilson, in California, in the decade following the end of World War I had revealed the immense extent of the Milky Way and the existence of extragalactic star systems. In 1931, at the time of Jansky's discovery, Hubble had used the 100-inch optical telescope on Mt. Wilson to penetrate several hundred million light-years into space. Within this region of space he had studied several hundred extragalactic nebulae—many of them galaxies similar to the Milky

Way, containing some one hundred billion stars. Furthermore, he had discovered that these galaxies were in rapid recession—and that at the limits of his penetration the recessional speeds were many thousands of miles per second and increased linearly with the distance of the nebulae. Hubble had estimated that the 100-inch telescope could penetrate 140 million light-years into space and that this volume of space contained two million extragalactic nebulae. It is understandable that, at the time of the discovery of radio waves from space, astronomers were skeptical that any further useful information could be obtained about the universe other than through the construction of larger and more efficient optical telescopes.

This belief existed when World War II ended, but soon a remarkable revolution occurred in the science of astronomy. Jansky and Reber had discovered that another window existed in the atmosphere, one through which radio waves from space could penetrate to the earth. This radio window exists at wavelengths a million times longer than those of the light to which our eyes are sensitive. Within a few years there had developed the new science of radio astronomy, in which the universe was studied at these long radio wavelengths. The critical discoveries that led to the awareness of the significance of the radio waves from space revealed that hitherto unsuspected processes of energy generation were prevalent in the universe.

In 1946 no one could have foreseen that Jodrell Bank would in time be deeply involved in these complex issues. However, at that moment in the summer of 1946 when we were faced with the problem of increasing the sensitivity of our radar equipment, I was, quite fortuitously, engaged in an observation with Banwell that formed the subject of the first scientific contribution ever made from Jodrell. The afternoon of 25 July was very hot, and just after four I suggested to Banwell that we discuss our problem in the relative coolness of the Park Royal. The receiving equipment of the meteor radar was switched on, and as we talked we faced the cathode-ray tube display. It was a simple range-amplitude display on which we normally observed the meteor echoes, and the gain, or sensitivity, of the receiver was adjusted so that the random noise could just be seen above the horizontal time base. After we had been talking for a few minutes, this noise level suddenly increased so that the whole vertical face of the tube was saturated. Our first instinct was to suspect that some fault had developed in the receiver. Almost as suddenly, though, the noise level decreased but remained abnormally high and fluctuating. In a few seconds we realized that we must be observing a solar radio outburst of the type that had caused great confusion during a crisis of the war in 1942.

This crisis was the escape on 12 February 1942 of the two German battle cruisers *Scharnhorst* and *Gneisenau* from Brest to their home ports in Germany. It was midday before the news of the escape reached the Admiralty, and the ships' relatively safe passage to Kiel at a time of great concern about Allied shipping losses was a grave blow to morale. Investigation revealed that the Germans had succeeded in jamming the British coastal defense radars. The Army Operational Research Group was immediately ordered to give priority to the investigation of this problem of the jamming of the British radars and to advise on antijamming measures. Two weeks later, on 27 and 28 February, another apparently severe case of jamming was reported. This time it concerned the anti-aircraft gun-laying radars working on a wavelength of 4.2 meters, but the expected enemy bombing attacks did not materialize.

In his analysis of this event Hey soon noticed that the reports of this jamming incident were confined to the daytime and that the operators' reports of the bearing of their aerials seemed to indicate a maximum effect in the direction of the sun. On inquiry to the Royal Greenwich Observatory he found that on 28 February a large sunspot group was on the central meridian of the sun. Hey's conclusion that the suspected jamming was actually radio emission, associated with the sunspot group, was issued in a secret report in 1942. In 1946, when Hey was able to publish the details of this event, he reported that the power emitted was equivalent to 10^5 times the blackbody radiation to be expected from the solar disk at a temperature of 6000 degrees Kelvin.

Both Jansky and Reber had anticipated that their cosmic static might arise from radio waves emitted by the sun, but neither could find a direct effect. Given the sensitivities of the equipment, no effect would be expected if the radiation was generated by thermal processes in a hot body at the 6000-degree temperature of the solar disk. The strange circumstance in February 1942 led Hey to the discovery that, when the sun was disturbed by large sunspots and flares, it became a powerful source of radio waves. When he published this conclusion in 1946, the radio experts were reluctant to believe that the sun was the source of this intense radiation.

The similar event accidentally observed by Banwell and myself on 25 July 1946 was another such outburst associated with an intense solar flare beginning at 4 P.M. on that day. The peak brilliance of the flare occurred twenty-seven minutes later, and at that time the intensity of the solar radio emission recorded in our trailer at Jodrell was over one hundred million times that to be expected from a body at the temperature of the solar disk.

Soon the investigation of the solar radio emission became a widespread activity in many parts of the world, but the subject never became a significant part of our own researches at Jodrell Bank. However, the news of Hey's conclusion about the 1942 solar outburst was the topic that led to Martin Ryle's initial research in Cambridge and to developments there that were soon to interact with our own work and form a coherent policy that gave an underlying strength to the development of radio astronomy in Britain.

Martin Ryle had graduated from Oxford in 1939 and proceeded to the Cavendish Laboratory to begin research work in the ionospheric group, led by J. A. Ratcliffe. At the outbreak of the war Ratcliffe joined TRE, and late in 1939 Ryle also joined TRE (then the Air Ministry Research Establishment). Six years later, as the war ended, Ryle returned to the ionospheric work at the Cavendish. This was not a success, but fortunately Ratcliffe, having had access to Hey's secret report on the *Scharnhorst* and *Gneisenau* incident, suggested to Ryle that he study the phenomenon of the emission of radio waves by the sun—in particular whether the sun was a radio emitter in the meter wave band when it was not disturbed by sunspots, and also to verify that the sunspots and flares were the actual sources of the enhanced radio emission. The aerials used by Hey had beam widths of the order of 10 degrees, and, since the solar disk subtended an angle of only 0.5 degrees and the spots and flares occupied only a small area of the disk, these conventional aerial systems were useless for the problem that faced Ryle. For example, to obtain a resolution of 0.5 degrees on a wavelength of one meter an aerial system with a diameter of 500 feet would be needed. It was this circumstance that led Ryle to the elegant solution of the radio interferometer. Two small aerials placed a distance apart on the ground and connected by cable to a common receiver could be used to obtain a resolving power equivalent to that of a single aerial with a diameter equal to the separation of the two aerials. It was the radio analogue of the optical interferometer, which had been used in the early years of the century on Mt. Wilson by A. A. Michelson and Francis Pease to measure the diameter of stars. With a student, D. D. Vonberg, he used this radio interferometer to measure the angular diameter of the solar region from which the radio waves originated and found it to be comparable to the area occupied by the sunspots.

Although these measurements of solar radio emission continued in Cambridge for some years, the interest of Ryle and his group soon shifted to the study of the cosmic radio waves. For entirely different reasons our own interests and research activities at Jodrell were soon to follow this course. In later years I appreciated how easy it would have been for Cambridge and Jodrell to be openly competitive in these activities. Fortunately for the survival of both centers of research this did not happen, because I had learned the hard way during the war that one did not advance one's own case by criticizing the other man—and so had Martin Ryle. In the give-and-take of the years ahead this lesson of wartime was often to be exhibited in the struggle for the increasing sums of money that we both needed to pursue our researches with the common aim of understanding the nature of the universe.

With this background I must return to the 218-foot transit telescope that we had first used before the end of 1947. With the transmitter connected to this aerial, we could see nothing unusual—the occasional meteor echo but no new phenomenon of the type that could be associated with the echoes from the cosmic-ray showers for which we were searching. There was, however, a very striking phenomenon.

The aerial received the cosmic noise from a strip of sky in the zenith, and, as the rotation of the earth carried the beam of the telescope across the Milky Way, the noise level increased dramatically, to a sharp peak followed by minor fluctuations and a subsidiary peak. At this time the measurements of Hey and his colleagues with the receiver of the army radar equipment in Richmond Park were becoming available, but the transit telescope gave us a far narrower beam and greater sensitivity than Hey's small array.

Victor Hughes was a research student with a temporary grant to enable him to study for a higher degree, and after our preliminary negative results, with the transmitter connected to the transit telescope, I saw that there was little future in that activity as far as his higher degree was concerned. The cosmic noise was a different matter. Clearly without any intention of doing so, we had built an instrument with far greater definition and sensitivity than any system so far used in these studies, and early in 1948 I advised Hughes to concentrate on the study of the cosmic radio waves from the zenithal strip of sky. If any worthwhile measurements were to be made, it was essential to determine the polar diagram of the aerial. Our only hope was to place a moving source of radiation well above the telescope by some means and measure the strength of the signal received in the paraboloid.

The solution was, as often in time of difficulty during the last few years, to telephone our former colleagues in TRE. This soon produced an aircraft in which we mounted a small transmitter, a camera obscura near the perimeter of the telescope to make an accurate plot of the position of the aircraft, and a radio telephone so that we could instruct the pilot. It was now a relatively easy matter to place correlation marks on the track of the aircraft, as traced on the camera obscura table, and on the recorder chart displaying the strength of the signal received from the transmitter in the aircraft. After a hundred or so flights at 3,000 feet in various directions across the telescope, we soon had the contours for the polar diagram of the telescope.

In the light of future events there was nothing particularly spectacular about Hughes's results. He had made an accurate study of the radiation from the narrow strip of sky passing overhead at Jodrell Bank and was able to derive some characteristics of the radiation by comparison with the recent survey made by Hey and with the earlier results of Jansky and Reber. He presented his thesis in the autumn of 1949 and soon left us, eventually to pursue a distinguished career in Canada as associate professor of physics at Queen's University, Kingston, Ontario.

Today the theses of those who have taken their higher degrees at Jodrell Bank occupy nearly forty feet of shelf space in the library. There are nearly four hundred of them, and Hughes's thesis is near the beginning. I looked at it again when writing this because I wanted to see once more the copies of the chart that made me so desperately anxious to be able to move that narrow beam. Night after night, advancing by four minutes the great hump of the radiation from the Milky Way—but always the same. If only we could direct that beam to any point in the sky, we could make a new map of the heavens. That was the recurrent theme and eventually the determination that led to our destiny.

The Arrival of Hanbury Brown

At the time in 1949 when Hughes was completing his measurements, I had to face the problem of the future use of the transit telescope. We had built the instrument in the hope that it would give us the extra sensitivity for the study of the radar echoes from the large cosmic-ray showers. The early investigations with the transmitter connected to it, late in 1947, gave little hope that we would achieve success.

One afternoon early in May of 1949 F. C. Williams, the professor of electrical engineering, telephoned. His call was to have an important effect on our future. I had known Freddie Williams since the time I first came to Manchester, in 1936. He was then an assistant lecturer in the electrotechnics department—a few steps along the corridor from the physics department. He had left to join the Air Ministry Research Establishment a year before the war began, and subsequently in TRE I saw him regularly when he was working on the IFF (Identification of Friend or Foe) equipment. At the end of the war he had returned to Manchester and at the time of his call was engaged in his brilliant work on early digital computers. However, his telephone call had nothing to do with this work. He asked me if I remembered Robert Hanbury Brown, with whom we had worked at various stages of the war. Of course I did; although he had gone to America in 1942, he returned to England after the war and joined the consulting firm of Sir Robert Watson Watt and Partners Ltd. The message Williams conveyed to me was that he had received a letter from Hanbury Brown, who wished to explore the possibility of returning temporarily to a university to do research for a Ph.D. He generously suggested that my developments at Jodrell might be of more interest to Hanbury Brown than his own work. The suggestion seemed like a gift from heaven. At that time Hanbury Brown was one of the most experienced electronic engineers in the country. Before and during the war he had been associated intimately with the development of airborne radar systems, first in England and then in America, at the Naval Research Laboratory, in Washington. A few weeks later he came to see me, and it was soon clear that the only problem was a financial one. With Blackett's help that was soon resolved, and by September 1949 he was occupied with the further use of the transit telescope for the investigation of cosmic radio waves.

It would be difficult for me to exaggerate the significance of his arrival. His wide experience of the world and his scientific and technical authority made an instant impression on Blackett and all members of the scientific establishment with whom we had contact. Blackett, the senior members of the university administration, and the external sources from whom we were demanding ever increasing sums of money no longer had to base their judgment solely on my opinion of what should be done. If they queried my extravagant demands, they could ask Hanbury Brown, and we soon began to make the kind of progress that would otherwise have been impossible.

At first that progress concerned the use of the transit telescope. Within a year he overcame, at least partially, the severe limitations of the transit telescope and produced results of the greatest significance to this new science of radio astronomy. More than thirty years later Hanbury Brown was asked to give an account of the sequence of events that followed. He wrote this:

They had built, with their own hands, a fixed paraboloid with the astonishing size of 218 ft. The original purpose of this remarkable dish had also been to detect cosmic rays by radar, but by the time of my first visit to Jodrell in 1949 they had given up that programme and realized that, by a happy accident, they had built a powerful instrument with which it should be possible to study the radio emissions from space. Lovell suggested that I should join them and take charge of a "cosmic noise" programme which had already been started by Victor Hughes. With Reber's paper on "Cosmic Static" in mind I accepted his offer with enthusiasm and went to work at Jodrell Bank in September 1949. Shortly afterwards I was joined by a research student, Cyril Hazard, with whom I was to work for several years. . . . As a first step we chose to work at the highest frequency at which the antenna could be expected to perform reasonably well; we wanted the narrowest possible beam. To that end we removed the original primary feed of the paraboloid, which had been designed to work on 4.2m with a high power transmitter using open-wire feeders; in its place we substituted a primary feed designed for 1.89m and connected it to the laboratory by 300 ft of coaxial cable with a loss of about 2.7 db. I well remember soldering this coaxial cable to the primary feed while perched on the top of the 126 ft tower at the centre of the paraboloid; it was not a job which I would care to do again. . . .

After describing how they had built a sensitive and stable receiver for this wavelength, Hanbury Brown continued,

We were able to make most impressive records of the "cosmic noise" power received from the zenith within a 2° beam. I shall never forget the thrill of seeing those first records. They showed clearly the broad maximum as the galactic plane passed through the aerial beam; they also showed a few discrete sources or, as we called them in those days, radio stars.

Then, in the early months of 1950, Hanbury Brown and Hazard took a vital step—they decided they could direct the beam away from the zenith by tilting the mast.

We set up two theodolite stations east and west of the mast and, very gingerly, tried tilting the mast by making successive adjustments to the 18 guys. It was a very anxious business; the tilting had to be done in almost imperceptible stages so as not to kink the mast and it took Hazard and myself about two hours, running from guy to guy and shouting at each other, to tilt the beam through one beam-width (2°). It was quite good fun on a fine day, although it was a rather slow way of scanning the sky. The early days of any new science are apt to be laborious. The largest tilt at which we could persuade ourselves that the mast was safe and that the beam shape was still respectable was 15°. This meant that we could survey a strip of sky between declinations +38° and +68°. . . . In those days very little was known about cosmic radio noise and our paraboloid was larger and had a much narrower beam than any other single antenna that had been used to scan the sky. Our programme, therefore, was simple and exciting: it was to look at everything in our field of view.*

In order to understand the drama that followed, it is necessary to describe the progress that had been made by Ryle in Cambridge following his use of the radio interferometer to study the sunspot radiation. In 1946 Ryle had been joined by Graham Smith, who, immediately after the war, had returned to Cambridge to complete his degree course. After working for some time on the sunspot radiation, they used the radio interferometer to investigate the fluctuating radio source that had been discovered by Hey in the cosmic radiation.

At the end of the European war in 1945 no further immediate tasks were in view for the Army Operational Research Group, in which Hey was working. Many members dispersed to peacetime occupations, but Hey remained in the establishment and, while awaiting further developments, seized the opportunity of studying the phenomenon of the cosmic noise that had limited the sensitivity of his anti-aircraft radars. Able to map the strength of the signal from different areas of the sky, he produced the first detailed radio map of the sky in the meter wave band and accidentally made a discovery of great significance. Although the signal strength varied from different parts of the sky, the signal was always steady, except from the direction of Cygnus. In this direction the intensity of the received signal fluctuated with periods of

*R. Hanbury Brown, "Paraboloids, Galaxies, and Stars: Memories of Jodrell Bank," in W. T. Sullivan, III, ed., *The Early Years of Radio Astronomy* (London: Cambridge University Press, 1984) p. 213ff.

a few seconds. His conclusion that a discrete fluctuating source of radio emission must be present in the Milky Way was to have far-reaching consequences for the future of radio astronomy.

In his account of this work Hey had drawn attention to the similarity between the fluctuations of the radio emission from this radio source in Cygnus and those of the sunspot and solar flare radio emission. It was therefore a natural step for Ryle and Smith to use their solar radio interferometer to investigate the cosmic source discovered by Hey. They made their first observations in 1948, and on the first night's record the interferometric pattern from the Cygnus source appeared as expected, but in addition there was an interferometer pattern separated by three hours from the Cygnus source. They had discovered a second localized source of radio emission in the constellation of Cassiopeia.

Simultaneously in Sydney, J. G. Bolton and G. J. Stanley had investigated the Cygnus radio source with a different type of interferometer. With this system Bolton and Stanley confirmed the existence of the localized radio source in Cygnus and discovered three others. One of these was coincident with the Crab nebula (the remnant of the supernova explosion observed by the Chinese astronomers in A.D. 1054), but neither the Cygnus radio source nor the other two could be positively identified with any of the objects on the star maps of the conventional optical telescopes.

Both in Sydney and in Cambridge these investigations were soon extended by the use of radio interferometers of greater sensitivity. In 1950, when Hanbury Brown and Hazard commenced using the transit telescope, Ryle and his colleagues in Cambridge had discovered fifty of these discrete radio sources, and the Australians had published a catalog of twenty-two similar radio sources. The remarkable fact was that, apart from the identification of one of the radio sources with the Crab nebula, not one of the discrete, localized radio sources could be positively identified with any of the stars or galaxies in the catalog of the optical astronomers. The distribution of these radio sources appeared to be isotropic, and the Cambridge group argued that they must be a hitherto undiscovered type of dark "star" in the solar vicinity. They became generally known as radio stars, and at that stage in 1950 the firm belief was that the phenomenon of the cosmic radio waves was restricted to the Milky Way system.

It was our good fortune that the nearest extragalactic nebula similar to the Milky Way—the M31 spiral nebula in Andromeda—was within the zenithal strip of sky that could be mapped by tilting the mast to the extreme limits. An elementary calculation suggested that, if our own galaxy were to be placed at the distance of the Andromeda galaxy, it should be detectable by means of the transit telescope. At that stage in 1950, when astronomers almost without exception believed that the cosmic radio waves so far studied were generated in our own galaxy, this appeared as a crucial question. Was the local galaxy unique in its emission of radio waves, or did other, similar galaxies behave in the same way? So far all attempts to identify specific extragalactic objects as radio emitters had been unsuccessful.

Beginning in August 1950 Hanbury Brown and Hazard began a survey of M31. When they were satisfied with the record from one narrow, 2-degree strip of sky, the arduous process of changing the tilt of the mast had to be carried out. They were nearly driven to despair by a sequence of thunderstorms in the middle of the night. The rest of us at Jodrell watched with incredulity as the evidence for radio emission from the nebula began to emerge. It took ninety nights to make the contour maps, and the survey was completed just before the vital region of the sky moved out of the radio-quiet night period when in transit. The results established beyond doubt that the M31 spiral galaxy in Andromeda also emitted radio waves of much the same intensity as the local Milky Way galaxy.

A brief and preliminary account of this work was published in *Nature* on 25 November 1950. It was the forty-first scientific paper published from Jodrell Bank; in astronomical importance and in significance for our developing plans, this publication exceeded by far anything that had previously emerged from the equipment in the Cheshire fields. By the standards of the day the work was a tour de force. Radio waves had been detected from a galaxy two million light-years distant, and it was no longer possible to regard the local galaxy as in any way unique as a radio emitter. However, in a curious way the observations deepened the mystery about the identity of the localized sources already cataloged by the groups in Sydney and Cambridge. Many of these sources were strong radio emitters—they had been discovered by means of small aerial systems of low gain. But even with the great sensitivity of the transit telescope the radiation from Andromeda was difficult to detect, for it was a weak source of radio emission compared with the majority of discrete sources already discovered. For this reason Hanbury Brown and Hazard reached the conclusion that these unidentified sources could not be extragalactic. They were too strong and must, therefore, be "radio stars" in the local galaxy.

On 12 October 1951, a year after Hanbury Brown and Hazard were completing their measurements on M31, both Hanbury Brown and Ryle spoke at a meeting of the Royal Astronomical Society. The former had been continuing the detailed survey of the zenithal strip with the transit telescope and had discovered several more of the localized sources of radio emission. Ryle described his failure to correlate any of the fifty sources cataloged by Cambridge with extragalactic nebulae. There was almost unanimous agreement that the localized radio sources were unseen radio stars in the local galaxy. Not quite unanimous. Tommy Gold, then a member of the Royal Greenwich Observatory, at Herstmonceux, and subsequently professor of astronomy in Cornell University, rose and argued vehemently that the sources must be extragalactic. He refused to believe that the difficulty of the intensity of the sources weighed against the extragalactic hypothesis. It was the beginning of a long and bitter argument with Ryle, which in the same room four years later was to be resolved in favor of Gold.

11

The Moon and the Aurora Borealis

In the summer of 1950, when Hanbury Brown and Hazard were making their measurements of the radio emission from the Andromeda nebula, I could not possibly have foreseen that eventually the study of the radio waves from the universe would completely dominate the research at Jodrell Bank. This came about through a complex of accidents and false trails—a salutary warning to those who believe it is possible to decide around a committee table what research should be financed.

Although our researches on meteors and the radio waves from space were consequences of my original plan to detect the cosmic-ray ionization, one item that I have not yet mentioned did form a part of my initial plan. This concerns the moon. At some stage during the war in my occasional contacts with Blackett, I must have mentioned the radar detection of the moon as a topic for postwar research. The evidence is in a letter I received from Blackett at the end of February 1945. His letter expressed the hope that I would find it possible to return to Manchester after the end of the war. He concluded, "We must meet again soon to discuss future cosmic ray experiments. . . . The radio detection of big showers and the distance to the Moon must be fully thought out." Now that the history of the attempts to obtain radar echoes from the moon has been studied, it is clear that scientists in several countries realized that the development of powerful radar systems during the war would make this possible. It is unlikely that I made any serious calculations about the problem until after the war, but an unpublished paper of May 1946, "Note on immediate prospects of obtaining radar responses from the Moon and the possibility of responses from the Sun and the Planets," reveals that I had intended to borrow radar equipment operating in the centimeter wave band to do this experiment.

We soon became engrossed in the meteor studies, and I have no recollection that we ever attempted a straightforward moon radar experiment with that unsophisticated equipment. One of the young recruits from my group in TRE arrived in June 1946 and was given the task of developing high-sensitivity equipment using long-duration radio pulses and a narrow-bandwidth receiver to study the lunar echoes. It must have been discouraging to be presented with such a formidable task in the somewhat crude environment of the Jodrell trailers. Clearly I still had to learn that what could be accomplished in the wartime environment of TRE could not easily be paralleled under the 1946 conditions at Jodrell Bank. Indeed, the task was a very hard one with the facilities we had available, and it was not until 18 July 1949 that we achieved success with a long-pulse 4.2-meter transmitter and a narrow-bandwidth receiver connected to the searchlight aerial.

The First Radar Contacts with the Moon

We had long since lost in any semblance of a race to be the first to obtain radar echoes from the moon. We came in fourth. Remarkably the first two successes were achieved within a few weeks in America and Hungary. Unknown to us, John H. De Witt, the chief engineer of an American broadcasting station, had made preliminary attempts to reflect radio waves from the moon in 1940 by using the transmitter of a New York radio station. These attempts did not succeed, and by 1942 De Witt was engulfed in radar research for the American army. After VJ-day, in August 1945, having several months still to serve as a lieutenant colonel in the army, he seized this opportunity of returning to the moon problem by assembling a powerful radar system under the guise of an official directive to develop a system for the detection and control of guided missiles. His "Project Diana" evolved in this way by means of U.S. wartime radar equipment on a wavelength of 2.6 meters. After several months of failure following the first trial with the system, in September 1945, success came on 10 January 1946. De Witt and his colleagues on Project Diana soon returned to private life. Although the project was an engineering success, there were few results of value from this first detection of radar echoes from the moon.

Compared with the resources of the U.S. Army Signal Corps the facilities in Hungary in this immediate postwar period were negligible, and yet on 6 February—only two weeks after the release of the news of the success of Project Diana on 24 January—news arrived that Hungarian scientists had succeeded in detecting the moon by radar. The man responsible was Zoltán L. Bay. Thirty-five years later when I was lecturing in the Franklin Institute in Philadelphia, Bay, then eighty years old, introduced himself to me, and only then did I get a glimpse of the strange circumstances that led to this extraordinary Hungarian success.

When Germany attacked the Soviet Union in the summer of 1941, Bay was a professor in the Budapest Institute of Technology. Hungary then became an ally of the Axis powers, and Bay was asked by the Hungarian government to develop a radar system for use against Soviet aircraft. Without any access to German radar developments he succeeded in producing a radar working on a wavelength of 2.5 meters, which went into service with the Hungarian army in 1944. By that time Bay had collected a considerable group of scientists and technicians, and he decided to develop a moon radar system on a wavelength of 50 centimeters. His laboratory then suffered a succession of disasters, first from Allied bombing raids and then from the invasion by the Soviet army and the siege of Budapest. By March 1945 he was able to make another start, but this time the Soviets dismantled and transported to the Soviet Union his laboratory and the factory manufacturing the Hungarian radar equipment. In August of 1945 Bay made his fourth attempt to construct a moon radar system,

179

and this time he succeeded—but in a most unconventional manner. He had no means of constructing a sophisticated receiver and display system such as that used in Project Diana. He solved the problem of integrating the reflected signals for a long period by displaying the receiver output on a battery of ten water voltameters. Most of the voltameters received receiver noise only, but one also received and integrated the reflected radar signal from the moon. On 6 February 1946 one of the voltameters showed a significant excess of hydrogen over the other nine after receiving the signals reflected from the moon for thirty minutes. By any standards this was a remarkable result of persistence in the face of most extraordinary hazards, but—like Project Diana—the experiment had little scientific value.

When Bay approached me at a reception before my lecture in Philadelphia in 1981, he clearly did not expect me to remember his lunar experiment of thirty-five years earlier, but I never could forget the complete astonishment with which we heard the news that a scientist in the Hungary of 1946 had made radar contact with the moon. For those of us who thought the only means of displaying a radar signal was on a cathode-ray tube, the story of the water voltameters seemed a fantasy.

Although neither the American nor the Hungarian lunar radar contacts of early 1946 produced any significant scientific results, the American report that the moon echoes appeared to be variable in strength seems to have stimulated the work in Australia. There F. J. Kerr and C. A. Shain had worked on radar developments during the war in the Commonwealth Scientific and Industrial Research Organization (CSIRO) radiophysics laboratory, and in April 1947 they decided to investigate the erratic behavior in the strength of the lunar radar echoes found in Project Diana. They enlisted the Australian government's shortwave broadcasting transmitter "Radio Australia." For three hours every morning they arranged to use the transmitter normally employed for transmitting to North America. This was at Shepparton, in Victoria, six hundred kilometers from their receiving site near Sydney. The transmitter, working on a wavelength of 15 meters, was pulsed by signals sent over a landline from this receiving site. The hindrance was that the moon passed through the directional aerial beam of the transmitter on only twenty-eight days of the year during the three-hour period when the transmitter was available. Even so, commencing in November 1947, the Australians were the first to achieve scientific results of value in the moon radar experiments. They distinguished two types of fading in the strength of the lunar echoes—one type was a rapid fading with a period of seconds, but this fading was superimposed on a much longer fading period of about thirty minutes. Kerr and Shain suggested that the short-period fading might arise from the apparent swaying effect of the moon known as libration. This effect arises because the moon's distance from the earth is not constant, and hence its orbital speed varies. On the other hand the rate of axial spin remains constant, and hence the moon does not always present exactly the same face to an observer on earth. There are a number of components of this libration, which has the effect that only three-sevenths and not one-half of the moon's surface is permanently turned away from the earth. In the radar case this would mean that the individual scattering centers on the moon are constantly changing, and it was this effect that Kerr and Shain suggested might be the cause of the rapid fading in the strength of the received echoes. They thought that the longer-period fading might have an ionospheric origin, but after the initial experiments with the Radio Australia transmitter the moon radar work was abandoned in Australia.

That was the international situation relating to the lunar radar experiments when we eventually succeeded in obtaining radar echoes from the moon at Jodrell Bank in July 1949.

The Development of the Lunar Radar Research at Jodrell Bank

During the early summer of 1953—that is, four years after our first success—the special lunar radar equipment developed at Jodrell Bank began to achieve significant scientific results. Long pulses of 30 milliseconds' duration were transmitted every 2 seconds on a wavelength of 2.5 meters. After a delay that varied between 2.4 and 2.7 seconds (depending on the moon's position in its orbit) strong echoes were received. A unique high-gain aerial had been built, consisting of an echelon of dipoles backed by a reflecting screen tilted back at 45 degrees. By adjustments to the aerial elements the moon could be observed for thirty minutes before and after transit for about two weeks every month. During October and November 1953 fifty thousand lunar echoes were photographed on the cathode-ray tube, and the analysis of these revealed that significant information could be obtained about the nature of the reflection process at the moon's surface. For the next eighteen months a mass of data on the nature of the fading was accumulated.

The investigation of the fading of the lunar radar echoes made at Jodrell Bank during the next few years provides an interesting example of how scientists progress in a research project. The problem was that the first lunar radar contacts in Project Diana revealed that the radar echoes received back on earth were not constant in strength, and the Australian observations indicated that there were two types of fading—a short-period fading (seconds) superimposed on a much longer fading period of the order of thirty minutes. In 1953 when, at last, a satisfactory moon radar system had been built at Jodrell, it became possible to study these fading effects systematically. In the first group of observations, carried out between September 1953 and August 1955, tens of thousands of the received echo pulses were analyzed. From these it was possible to give conclusive proof that this short-period fading was indeed due to the libration of the moon, as had originally been suggested by the Australians. The rate at which the moon librates varies with its position in the orbit around the earth. Thus, if the fading was due to the libration, it should be possible to discover a relation between the short-period fading and the position of the moon in its orbit. By measuring the correlation in the strength of successive echoes, one could determine if this correlation depended on the moon's orbital position. This correlation was found to vary with the position of the moon and therefore left no doubt that libration was the cause. However, the correlation was only one-half of that to be expected. When these results were communicated for publication toward the end of 1955, the reason for this lack of correlation was unknown, but on the whole it was concluded that the nature of this fading was consistent with the assumption that the moon reflected radio waves in such a way that the disk appears uniformly "bright" (as it does for the reflected light of the sun by which we see the moon—that is, apart from the shadows cast by the lunar mountains).

This conclusion seemed satisfactory, if rather dull, and since the small group that had carried out this work had completed its studies for higher degrees its departure could well have signaled the end of our interest in the moon. However, I did not like that loose end—why was the correlation between the strength of successive echoes only half of what it should have been on the basis of that predicted from the variation of the effect of libration as the moon moved in its orbit? At that time I had in the undergraduate class an outstanding young man, John V. Evans, who was keen to work for a higher degree at Jodrell. In due course his immediate ambition was fulfilled, a grant was obtained, and Evans appeared on a motorcycle at Jodrell Bank in October 1954. He was eventually attracted to the United States, where he achieved higher and higher positions; when I last had contact with him, in the early 1980s, he was a vice president of the Comsat organization.

Evans soon settled the problem of the rapid fading. He argued that, if the moon was reflecting the radar pulses as a uniformly bright scatterer, it should be possible to find the effect by studying the shape of the individual echo pulse. There would in this case be a significant time delay between the part of the incident pulse scattered from the nearest point of the moon (the center of the disk) and that scattered from the limbs—about twelve one-thousandths of a second—that is, the time difference for the radar pulse to be scattered from the front of the disk and the limbs would be equivalent to twice the lunar radius of 1,738 kilometers. He found that the shape of the pulse was not changed and that an accurate range measurement was consistent with the concept that the pulse received back on earth had been scattered from only a small area around the front face of the moon. He soon reached the remarkable conclusion that for radar waves the moon reflected in quite a different manner than for light waves. In the latter case the disk is uniformly bright, but in the radar case it is limb dark—the radar echo was returned from a region at the center of the visible disk, which had a radius of only one-third of the lunar radius. Everything now fell into place, not only the problem of the correlation level between successive echoes was solved but also the puzzling fact that for the radar waves the reflection coefficient of the moon seemed to be only one-tenth of that which had been calculated on the assumption that the whole visible surface was equally effective in scattering the radio waves. In his conclusion to the scientific paper on this work Evans wrote, "Since the effective depth of the Moon is 1 millisecond or less it becomes possible to use the Moon in a communication circuit with modulation frequencies up to 1000 cycles per second. This is probably just sufficient for intelligible speech and could be used for teletype."[*] A remarkable conclusion that would have been of vital importance had it not been for the development of communication by earth satellites.

In the meantime the problem of the longer-period fading had also been solved.

One Sunday afternoon early in 1954 I answered the telephone. It was W. A. S. Murray, one of the students then working on the moon radar. Could he talk to me urgently about his work? He came, full of enthusiasm, because, he said, he had reached the conclusion that the ionosphere was responsible for the long-period fading—but not in the way the Australians had suggested. The phenomenon, Murray argued, was caused by the "Faraday effect" as the radar waves passed through the ionosphere. That is, the plane of polarization of the radar waves was being rotated as they traversed the ionized regions in the presence of the earth's magnetic field. This Faraday effect is a well-known phenomenon to physicists, and the rate of rotation depends on the electron density, the magnetic field, and the frequency of the radio waves. Now, in all the moon radar observations made so far, the transmitted wave had been plane polarized—that is, the primary feed of the aerial systems had been either horizontal or vertical dipoles. For example, if the transmitted wave was in the horizontal plane and if in the transmission to the moon and back the plane of

---

[*]J. V. Evans, "The Scattering of Radio Waves by the Moon," *Proceedings of the Physical Society B (London)* 70 (1957):1112.

polarization had been rotated through 90 degrees, the strength of the signal received with the same horizontal aerial would be very low. On the other hand if arrangements could be made to receive the scattered signal with a vertical dipole aerial, the maximum signal strength would be obtained. I was never able to discover whether Murray had produced this idea himself. He certainly claimed to have done so, but others at Jodrell also claimed that they had suggested that the Faraday effect might be responsible for the fading. In any event, on that Sunday afternoon Murray was able to convince me that the calculations of the rate of rotation of the plane of polarization were compatible with the observed rate of the long-period fading. Furthermore, he proposed a series of observations by which he could prove that this was the case.

A few hundred yards distant from the hut containing the moon radar there was a small steerable paraboloid 25 feet in diameter that I had reclaimed from use as a demonstration exhibit at the 1951 Festival of Britain, marking the centenary of the Great Exhibition of 1851. The request made to me by Murray was that he be allowed to use this paraboloid. His proposal was to transmit to the moon with the existing horizontally polarized array but to receive the scattered signals on this paraboloid. He proposed to feed this paraboloid with a crossed dipole array so that he could receive on either the horizontal or the vertical polarization and switch between them at one-minute intervals. If his idea was correct, then as the signal strength decreased on the horizontally polarized dipole, it should increase on the vertically polarized system.

This was ideal for the moon observations proposed by Murray, and in five afternoon and evening sessions during the lunation of March 1954 he and another student, J. K. Hargreaves, transmitted on one polarization and switched at one-minute intervals between two polarizations using this paraboloid as the receiver. The results were conclusive—the long-period fading was caused by the rotation of the plane of polarization of the radio waves as they passed through the ionosphere. Within a few years John Evans, having brilliantly solved the rapid-fading problem, used this technique to make an entirely new series of measurements of the total electron content between the earth and the moon, but the full development of that idea belongs to a later period.

185

The Aurora Borealis

The moon radar work was the only item in our eventual researches that I had originally suggested as a research topic to Blackett toward the end of the war. Our radio echo studies of meteors developed as an unexpected result of my intention to use the army anti-aircraft radars for the detection of the ionization from large cosmic-ray showers, but there was yet another unexpected result of this work. This happened one night in August of 1947 when we were recording the echoes from the meteors of the Perseid shower. By that time we had seen many thousands of echoes from meteors on the cathode-ray tube. During the night of 15–16 August I was working in the trailer with Clegg and Ellyett when a few minutes after midnight a completely new phenomenon appeared on the cathode-ray tube. The Perseid meteors approached their maximum, and on the tube were a large number of echoes typical of the transient echoes from the meteor trails. Suddenly, ten minutes after midnight, a new type of echo appeared, extending over a considerable range of the time base. The echo had the appearance of much frothing and bubbling, as though the whole was composed of a multitude of individual scattering centers. This was not a transient echo. We looked at the tube in astonishment as the echo lasted for seconds, and then minutes, and moved to and fro along the time base (that is, in range). After a few minutes I decided to look at the sky, and outside the trailer a beautiful sight awaited me. The northern sky was ablaze with great auroral streamers—a rare occurrence so far south. I had never before seen an aurora borealis, but it was instantly recognizable. The brilliant streamers rose from the northern horizon nearly to our zenith. Frantically we made whatever measurements we could, and by good fortune we were observing the Perseid meteors with two radars on different wavelengths, with one connected to the steerable searchlight aerial. The echoes vanished from our cathode-ray tube after about twenty minutes, but we had been able to establish that the radar echoes coincided with the appearance of a faint blue-gray auroral cloud near the zenith at the tip of one of the streamers. By mid-September we had published an account of this episode, including the determination of the electron density in the auroral cloud. When Sir Edward Appleton read this paper, he complained that he and his colleagues had made similar measurements with the early ionospheric equipment they had operated at high latitudes during the International Geophysical Year of 1935. Sir Edward, a senior and influential member of the British scientific establishment, had already had a confrontation with J. S. Hey over priorities concerning the solar noise observations. It was my first glimpse of the ever present urge among scientists to claim priority. However, we were more concerned with our good fortune in opening another line of research, which for some years provided a fruitful subject for many of my students and gave important information about the nature of the aurora borealis.

The radar studies of the aurora soon became widespread in many laboratories, particularly those at high latitudes, and in recent years many millions of pounds have been spent on international projects in an attempt to understand the precise processes by which the occurrence of the aurorae are linked to solar eruptions and events in interplanetary space. But that August night in 1947 contained something new, and, like the sky outside, it was spectacular and memorable.

The Strange Incident of A and X

By the early 1950s the successes with the radar studies of meteors, the moon, and the aurorae and the rapidly developing researches in radio astronomy with the transit telescope were evidence that these new techniques could make a unique contribution to scientific development. However, during this period when we were establishing ourselves as a research group, I was drawn into a bizarre and worrying episode that involved two young men whom I will refer to as A and X. Now, X arrived at Jodrell Bank in October 1950 at the same time as A. Although not involved in the same research program, he soon became involved with A in a curious manner. Both A and X were in their mid-twenties but had radically different backgrounds. X came to us from an overseas country and had made his way in life from humble origins by sheer scientific excellence. The young man A provided a contrast. His father was a distinguished naval person, and A had followed his father's career and had survived the harshest of physical duties while serving on convoy escorts during the war. After demobilization he came to the university, like many other young men at that time, to study for a degree and was one of my undergraduate students until he took his degree.

For some two years no particular incident occurred. Both A and X proceeded with their separate researches, and as far as I was aware their contacts were purely scientific ones in the normal course of the mutual help that the staff at Jodrell gave to one another. By the autumn of 1952 I was well aware that a variety of political and religious attitudes existed among the members of my Jodrell group. I had discovered, too, that X was a rather aggressive member of the Communist party. However, the charter of the university demanded religious and political freedom for its members, and I had been careful to keep prejudices one way or the other from becoming manifest at Jodrell Bank. This was the epoch when Senator Joseph McCarthy was running his campaign against the infiltration of Communists into public offices in the United States and was exploiting public concern about the spread of communism in Asia and Europe. In Cheshire I had moved into a community that was strongly Conservative in outlook, and unfortunately the growth of Jodrell had, for some reason, given the impression that we were the type of community that McCarthy was castigating. The impression, however it arose, was very far from the truth, and X was far outnumbered in the Jodrell group by those with opposite political and religious affiliations.

Nevertheless, I was soon to discover that, whatever the facts might be, the impression that we were a community of Communist sympathizers was widespread in the surrounding district. This had been brought home to me one wild and wet winter evening. My wife and I had moved with our family to the nearby village of Swettenham, a few miles from Jodrell Bank. We were still relative strangers to the community on the night in question when, because of some organizational lapse, she arrived at the local railway station of Goostrey with two young children in the darkness without means of traveling the several miles to our home in Swettenham. One other person alighted from the train, and my wife asked him, if by chance he was proceeding in the direction of Swettenham, could she beg a lift? "Hop in"—"Are you sure it's not out of your way?"—"I don't mind who I give a lift to except those b—— Communists at that place Jodrell." A moment's staggered amazement—and subsequently Colonel D, who turned out to be the driver of the car, said that in that moment of silence he realized he had committed a faux pas.

In any event, that was the background to my amazement when, during the last week of October 1952, I read in the *Manchester Guardian* that A and X, using the Jodrell Bank address, had been signatories to a telegram to President Truman demanding the release from custody of the New York couple Julius and Ethel Rosenberg, who, after an internationally famous trial, were found guilty of having passed atomic bomb secrets to Soviet agents. (The couple were executed in 1953.) My letter of 31 October to Vice-chancellor Stopford expressed my dismay:

> The unfortunate linkage of Jodrell Bank with the telegram to the United States in Thursday's issue of the Manchester Guardian may well have serious consequences for our work. . . .
>
> My own view of the situation is so grave that I feel impelled to ask if you would consider delivering a severe reprimand to the signatories of this telegram, and to request them to make a public apology in the Manchester Guardian for implicating the University in their private affairs.
>
> May I emphasise that this request is made entirely on non-political grounds. My deep concern is that very serious repercussions may follow for our work at Jodrell Bank, and my reaction would be the same whatever the political sympathies of the signatories to the telegram.

The vice-chancellor responded that the episode was "unfortunate and disgraceful" and that he had received letters of apology. But that was by no means the end of the affair, since further publicity arose from an open telegram delivered to Jodrell Bank from Emanuel Bloch, the Rosenbergs' defense lawyer, inviting X to testify at the trial with "all expenses paid." However, the most astonishing event still lay ahead.

Within a few days I had summoned A and X to my office and had left them in no doubt of my reaction to their involvement, particularly emphasizing that if they wished to engage in such politics, they would please do so from their private addresses and not involve their association with Jodrell Bank. By chance a few weeks earlier the distinguished naval person who was A's father had invited me to dinner with my wife. After such a pleasant acquaintance with A's background it was perhaps inevitable that as A and X were leaving my office I said, "As for you, A, I'm surprised that with your upbringing and background you should be associated with this business." He left the office with X but a few minutes later pushed open the door unceremoniously and said, as though I was the student culprit, "Look here, if you're going to speak to me like that, I must ask you to receive a visitor from London."

Almost immediately a letter from a certain room in the War Office in London was placed in my hands. It was from a Major K, who desired to see me in connection "with a private interview which you had recently with Mr. A of your establishment." He came: I asked to see his identity pass. It was a short interview. "Just be careful how you treat Mr. A," I was told, "He's one of the best agents we have in the country." A had penetrated a Communist cell in the northwest of England and was to be left unmolested to carry on his valuable work. Writing about this event thirty-five years later, I still recall my feeling of utter disbelief as Major K departed—the strange uncertainty of where one stood in this affair, of who was spying on whom, and whether X himself was suspicious of A? I never found out, but at that moment in November 1952 I had one urge—and that was to get rid of both A and X at the earliest opportunity—an ambition I did not achieve for another two years. The episode of A and X in the autumn of 1952 was the first, and I hoped the last, encounter I would have with the hazy world of espionage and counterespionage—but I was still innocent of the international strains and stresses of the postwar world.

12

The Dream of the Telescope

I have already explained how we first used the transit telescope as a radar instrument to search for cosmic-ray echoes toward the end of 1947. The results were negative, and Victor Hughes then used the telescope to study the galactic radio emission in the zenithal strip of sky. By the time he had finished his measurements, in March 1949, a small number of localized (and as yet unidentified) radio sources had been discovered by use of the radio interferometers in Cambridge and Sydney. Interest in the radio emissions from space was increasing rapidly. During that winter of 1948–49 I had begun to talk to Blackett about the desirability of building a large radio telescope that could be mounted in such a way that it could be directed to any part of the sky. The primary reason for my presence at Jodrell Bank was still the attempt to detect the large cosmic-ray showers by radar, and I argued that our experience with the echoes from meteor trails showed the importance of the orientation of the beam with respect to the direction of the column of ionization. We believed then that the sensitivity of the transit telescope was near the required limit, and so I held that it could well be that certain orientations of the beam would give a greater chance of success than the present, vertical direction.

191

With Hughes's charts before me I could now demonstrate that this very large paraboloid worked as we had intended. With this evidence the immense value of such a steerable narrow-beam aerial system for the exploration of the radio waves from space became manifest. It is a very great tribute to my memory of Blackett that I do not recall that he ever argued against the idea or discouraged me in any way. Indeed, he would have had ample justification for making me suppress any further thought of such a project. On the contrary, he was excited by this prospect of exploring the heavens and said that, if I could find someone who could build such an instrument, he would do his best to help. There were to be many heated exchanges between us during the next few years, but he never once wavered in this promise of help.

The Search for an Engineering Solution

At that time I clearly had no idea of the formidable engineering task that I was proposing. We had built a 218-foot paraboloid ourselves, anchored to the ground in tons of concrete. I could visualize it suspended in some way so that we could swing the bowl of the telescope in any direction we wished. During the war I had caused large rotating paraboloids to be mounted under the belly of bomber aircraft flying at 20,000 to 30,000 feet, and it seemed to me that the engineers who had carried out that work could easily tackle the much larger ground-based steerable paraboloid that I was proposing now.

My expectation of an enthusiastic response from these contacts was not fulfilled. Indeed, by July 1949 I had encountered severe discouragement from every large engineering firm in the country that I thought might be interested. The responses took two forms: either the firm was too busy with work of national importance or the engineers had looked at my proposal and concluded that it was not technically feasible. On reflection in later years I am astonished that with such massive engineering discouragement I did not abandon the idea. However, some quirk of my temperament resented these dismissals of an idea. It had happened early in 1942 when Handley Page said that his Halifax bombers would not fly with the H₂S scanner underneath the fuselage. Then I could bring superior authorities to bear on the firm, but in 1949 I was alone and temporarily devoid of ideas of how to proceed.

The vacuum did not last for long, and before many months had elapsed I had not merely one but two entirely separate developments of the idea in progress. These proceeded independently, with neither party aware that I was also dealing with another firm. The consequences were almost disastrous.

The Special Projects Department of Head Wrightson

In July, when my direct approaches to the engineering firms had been rebuffed, I had the idea of seeking advice from George M. Sisson, the head of that part of the Grubb Parsons firm in Newcastle on Tyne, which had built many successful optical telescopes that were working in many parts of the world. Sisson did not think that his own firm of Grubb Parsons would be willing to design and build the telescope, but he placed me in contact with Head Wrightson & Aldean Ltd., a firm specializing in structural aluminum work. The special projects department of that firm soon produced an elegant sketch of a "200ft radar paraboloid screen." This seemed to me to be exactly the kind of engineering version we had expected, and when I showed it to Blackett he became enthusiastic and thought the drawing was extremely beautiful, but difficulties appeared immediately. We wanted to know how much the telescope would cost, and the sales manager, W. H. Ryan, informed me that the firm would need a fee of £900 to carry out further design work in order to provide an estimate of cost to within 10 percent. Blackett wanted to get the university involved officially in the project, and he persuaded the university council to vote this sum. Blackett asked me to discuss the arrangements with the university bursar, R. A. Rainford, who, in his wisdom, simply said that the university would pay the money when the firm had supplied the requisite information. The university never parted with that £900.

A second difficulty was my demand that the size be increased from the 200-foot aperture of the sketch. I had asked for a paraboloid at least as large as the 218 feet of the transit telescope and had freely talked about a diameter of 250 feet as a target. In my letter of 5 April 1950 to F. J. Dean, the managing director of this special projects department, I thus confirmed, "We desire a paraboloid of diameter 250 ft. If the cost of this is liable to be very high, then it would be most useful to have approximate estimates if the size were reduced to, say, 225 ft or 200 ft."

By the end of April 1950 our original idea that we might build this telescope for £50,000 was beginning to look absurd. Ryan was already talking about at least £150,000 for a 200-foot paraboloid, and for my part I was beginning to make suggestions for maintaining the size but reducing the cost (such as relaxing the wind-loading conditions) that were to become a common refrain in the next few years. I had kept Blackett in close touch with these developments, and he made it clear to me that he thought a sum of £150,000–£200,000 for this telescope would be an impossible target. I thought it was time to persuade Blackett to come to Slough with me to see the model and have a firsthand discussion with Dean and his colleagues about the project.

Various factors contributed to the disastrous nature of that meeting on 11 May, a principal one being that I had not received a letter from Dean in which he summarized the six different schemes on which we had been engaged. These plans had cost £1,280 (nearly a third more than the university had agreed to pay) exclusive of the model of the telescope on Dean's desk, which he said could not possibly be built for less than a quarter of a million pounds. This was already five times the amount I had originally suggested to Blackett less than a year earlier.

Blackett reacted with increasing hostility to Dean and his various suggestions for reducing the cost and soon made it clear that he did not think I was competent to deal with issues like this and that I was wrecking the whole scheme by demanding such a large instrument. In fact, before our interview with Dean terminated, Blackett had demanded that he produce details of the cost of a telescope with an aperture of 200 feet and meet us again at Jodrell as soon as possible.

Dean never arrived for that meeting arranged for 27 June 1950 at Jodrell. With Blackett impatiently awaiting his arrival my efforts to locate Dean eventually produced the information from Head Wrightson that their special projects department had been closed on 25 June, that Dean was no longer a member of their organization, and that I should have received a letter informing me of this fact. I never received that letter. My subsequent exchanges with the technical director (F. J. Walker) of the main Head Wrightson organization were completely unsatisfactory, and no explanation of these strange happenings was ever given. On 18 July I wrote to Walker finally terminating our arrangement.

Naturally these bizarre occurrences confirmed Blackett in his opinion of the Head Wrightson engineers and of my capacity for dealing with such issues. He was certainly a very angry man as he stormed away from Jodrell on that day in June, but in another direction our fortunes were less calamitous.

My Meeting with H. C. Husband

In chapter 9 I explained the circumstances that led me to make contact with the firm of Coubro & Scrutton in 1947 in connection with the 126-foot tubular steel mast for the transit telescope. In the early months of 1949, when I was making my direct approaches to the engineering firms, I wrote on 15 February to F. D. Roberts, my contact in Coubro & Scrutton.

As I quickly learned from Roberts, his firm had decided that it could not tackle the project, and he advised me to approach "one of the leading structural firms, such as Dorman Long or Vickers Armstrong." Having already been rebuffed in these directions, I again pressed Roberts for help. On 21 July he said his company had business connections with a firm of consulting engineers who might be able to help. After some months, and after Sisson had already placed me in touch with Head Wrightson, Roberts brought H. C. Husband to Jodrell. We met on the afternoon of 8 September 1949 near the perimeter of the 218-foot transit telescope. Husband asked me about my problem. I explained that I wanted to build a telescope of at least that size, mounted so that it could be directed to any part of the sky, but that several engi-

195

neering firms had said this would be impossible. Husband surveyed the structure and without hesitation replied that he did not think it was impossible—"about the same problem as throwing a swing bridge across the Thames at Westminster."

In due course I found out that Husband was an expert on bridge design and that he was the senior partner in the firm of Husband & Co., consulting engineers, with main offices in Sheffield and London. For almost the whole of the remainder of my career, my work was to be entwined with Husband and his firm; however, this close association did not follow immediately from that autumn afternoon in 1949. Roberts had brought Husband to Jodrell, and it was he who continued the immediate contacts. At the end of November, Roberts wrote, "Such information as I have, suggests . . . that a very rough preliminary estimate would be about £45,000." I pressed Roberts for more information and a sketch, and on 20 February 1950 he posted the sketches to me— but had no later estimates of the costs.

I had already received the Head Wrightson drawing. Writing about this thirty-six years later, I am very glad to have discovered both of these original drawings, for long buried in a forgotten file. Husband's sketch was not a beautiful drawing like the Head Wrightson one, but it was extremely practical—based on seventeen sheets of detailed drawings and calculations of the wind forces and stresses. Indeed, this Husband drawing dated January 1950 bears a remarkable similarity to the telescope that was eventually built. The Head Wrightson design in light alloy, although appearing more elegant in the sketch, presented the problem of the lightness of the structure and its consequent stability in high winds. We were to face plenty of wind problems with the Husband design, but at least the dead weight of the steel structure meant that there was no fundamental problem of overturning in high winds. In later years many people questioned the wisdom of building in steel instead of light alloy. Husband's response was always that in light alloy one would have to add redundant weight to withstand the wind forces and that the total weights of the structures would thus be similar.

By the end of May 1950 Roberts had concluded that we must negotiate with Husband directly. A meeting at Jodrell on 14 June of Husband, Blackett, and myself was entirely different from that unfortunate confrontation with the Head Wrightson engineer at Slough a month earlier. Blackett was impressed with Husband's proposals and with his businesslike approach to the problem. Husband produced the drawing I had already received through Roberts in February, and, as he confirmed in writing a day later, he thought that we "should be reasonably safe with a figure of £100,000," adding, "I would emphasise, however, that for anything as unusual as this a design in much more detail than we have prepared already is essential in order to obtain firm tenders for carrying out the scheme." Blackett was satisfied, and we applied immediately to the Department of Scientific and Industrial Research (DSIR) for a sum of £120,000 to cover the cost of the design and erection of the instrument. With remarkable speed at a meeting on 22 June the DSIR voted the agreed sum of £3,300 to enable Husband to undertake the detailed design so that tenders could be invited for the construction of the telescope.

Blackett's Strategy

The fact that we were able to make an immediate application to the DSIR, so that it agreed to vote the money for the design study a week after our meeting with Husband, was entirely due to Blackett. Now I was to witness the fruits of his long experience in peace and war in dealing with major projects. Blackett told me that it was essential to get the backing of the key scientists in the country and to determine how this would fit in with the other scientific developments in the United Kingdom, especially in astronomy. In the autumn of 1949 he suggested to the council of the Royal Astronomical Society (RAS) that, in view of the rapid development of the researches in radio astronomy, it might be advisable to convene a committee to consider what should be done to support the subject in the United Kingdom. I was perturbed because I thought the reason for proposing that such a committee be formed would be to obtain support for the telescope. Blackett agreed and said the only way to obtain such support was within the context of other proposals for developments.

There were only two other relevant groups in the country—Ryle's in Cambridge and Hey's in the Army Operational Research Group (AORG) organization—and after the RAS council had agreed to set up this committee, Blackett told me to prepare a paper in collaboration with Hey and Ryle. On 18 January 1950 I met Ryle and Hey and compiled a document that gave a "summary of a programme of work in radio astronomy which Great Britain should be able to carry out during the next 5 years." As an appendix I attached a paper giving details of a "programme for a large radio telescope which might be constructed in the near future." On 27 January, Ryle wrote that neither he nor Ratcliffe agreed with my paper, because it mixed up the broad outlines with the detailed accounts of the work in progress at the three centers. The committee had already arranged to meet in London on 10 February, and I agreed with Ryle's version, which separated the work and future programs of our three groups into appendixes.

Ryle's objection to my paper had the entirely favorable outcome that my case for the steerable telescope was the only document before the committee at its first meeting, on 10 February. We met in the RAS rooms in Burlington House.* After an explanation by Blackett of the possible function of the committee, there was a low-key discussion about arrangements for routine recording of solar and galactic noise, and for routine meteor recording. Ratcliffe raised the question of facilities for radio observations of the 1952 solar eclipse, and M. A. Ellison of the Royal Observatory, Edinburgh, advocated improvements in cooperation between solar and ionospheric workers.

When these discussions had run their course, the last minute (no. 11) of that meeting recorded,

---

*The council of the RAS had invited the following to serve on the committee: W. M. H. Greaves (the Astronomer Royal for Scotland, who acted as chairman), Blackett, Ratcliffe, Hey, Ryle, H. W. Newton, Prentice, Appleton, H. S. W. Massey, and myself. Only Appleton and Massey were absent from this first meeting. M. A. Ellison acted as secretary of the committee.

Dr. Lovell then asked for consideration of his project for building in this country the 200ft* paraboloid receiver so mounted that it could be swung round to any region of the sky. Preliminary drawings and a memorandum were submitted. After some discussion it was agreed that this was a matter of great importance for the future of radio astronomy in Britain and that the matter should be placed on the Agenda of the next meeting.

Appleton was not at this first meeting. He had recently left the DSIR to become the principal and vice-chancellor of the University of Edinburgh. Even so, like Blackett, his seniority and influence coupled with his particular interest in the subjects under discussion made it essential to involve him. That was the primary reason for convening the second meeting of the committee in Edinburgh. At the first meeting it was agreed also to invite the Astronomer Royal, Sir Harold Spencer Jones, to attend. These were the days in the United Kingdom when no major project in astronomy stood the slightest chance of success without the backing of the Astronomer Royal.

This second meeting of the committee was arranged for 27 February at 2:15 P.M. in "Sir Edward Appleton's Apartments, Edinburgh University," and neither Blackett nor I had any doubt that a negative reaction from the committee would mean the end of the project. Writing about this more than thirty-five years later, I still feel the tenseness, the moments of dejection, and the unhappiness that preceded that afternoon in Sir Edward's rooms. In the two weeks since the London meeting, I had been under relentless pressure from Blackett to make sure I could justify the scientific case. Then came the sheer physical discomfort of a Sunday night train journey to Edinburgh.

Blackett was occupied in so many diverse activities that he seemed to be traveling ceaselessly. There were not yet rapid internal airline links between the major cities, and he frequently explained to people that he lived and slept on British railways.

*The figure 200 feet appears in the minutes. My memorandum submitted at the meeting was headed "Notes on the projected 200–250ft aperture paraboloid." Husband's drawing shown at this meeting was for a 250-foot-diameter paraboloid. There had been dispute between Blackett and myself about the size. He argued that, by demanding a size of 250 feet, I would ruin the project and that it would be safer to go for a smaller telescope. Fortunately my view prevailed, and after this first meeting the aperture of 250 feet was always mentioned.

Perhaps he did sleep on that Sunday night in the bumpy sleeper carriage of those days. For me the journey was a sleepless nightmare. The night trains went to Glasgow, and passengers destined for Edinburgh were ejected at 6 A.M. at a junction called Carstairs. That morning of 27 February was icy. There was a hoarfrost and the kind of dirty fog that made one reluctant to breathe. Even if there had been a breakfast room at Carstairs, Blackett would have had no intention of using it, since he had arranged for us to have breakfast some hours later with a colleague in Edinburgh with whom he had some topics on cosmic rays to discuss. It was at Carstairs on that dimly lit platform while we waited for the Edinburgh train that Blackett raised the question of our "tactics" at the meeting. Naturally I had assumed that if the telescope was built, it would be at Jodrell. That was not in his tactical scheme. "Of course, you will emphasize to the committee that you are not putting forward this plan on behalf of yourself as something to be built at Jodrell, but as a project for the UK with the committee to decide the best place to build it." To my consternation Blackett thought it would be best to build it at the Royal Observatory at Herstmonceux.

There was no retreat. At last the slow journey to Edinburgh ended as dawn was breaking, and, after a seemingly endless morning with Blackett now interested in cosmic rays and political matters, we entered the spacious rooms in the university occupied by Appleton. Apart from himself and Greaves, as chairman, the Astronomer Royal (Spencer Jones), Blackett, Hey, and Ellison, as secretary, were present, but these were the key people who within a few hours would vote on whether the project should proceed.

In retrospect it seems extraordinary that neither Ryle nor Ratcliffe came to this meeting, which was to be of cardinal importance to the development of radio astronomy in the United Kingdom. In fact, Ryle had by this time produced his memorandum, but it was instantly agreed that any discussion on it should be postponed to the next meeting, together with the paper produced by Hey. It was my good fortune that the papers by Ryle and Hey highlighted the strength of my own case for the large telescope. Neither he nor Hey had any immediate equivalent proposals. Ryle was in the process of developing the interferometric techniques. Still believing that the localized sources were within the local galaxy, he wrote, "Further exploration of the galaxy will therefore require the use of an aerial system having an appreciably greater aperture than the present one (6000 square feet). No plans are at present in hand for the construction of such an aerial until more information is available on the optimum band of wavelengths for such an extension, and on the best type of aerial to be used." In his paper Hey gave a short account of his postwar work in AORG but pointed out that in his case "the pursuance of basic research beyond the stage where practical implications can be foreseen is strictly limited."

So it happened that almost immediately the chairman (minute 4) invited me "to put forward [my] proposals for the construction of a 250ft diameter steerable paraboloid (radio telescope) for use in radio astronomy." Then at last I began to see the master strategy of Blackett working its magic. When I was a schoolboy, one of my important examination papers asked almost precisely the questions the master had foreseen. It was like that now. Exactly as Blackett had foreseen, I was first challenged on the reasons and then on the technical problems. I had answered all the questions only hours before—at midnight on the greasy platform in Manchester, or in the predawn fog and ice at Carstairs. On the question of where it should be built, I reluctantly obeyed Blackett's instructions, but I never finished the sentences. With almost one voice the members of the committee interrupted—they wished to make it clear that, if any such device was to be built, a condition must be that it should be built at Jodrell Bank, where the techniques existed, and be run by the person who had proposed it.

In my experience votes are never taken in a committee where issues of great significance are concerned. The fact that the chairman did not call for a vote on my proposals heightened the dramatic quality of that afternoon. The chairman simply asked each member to state his final opinion on whether the committee should back the proposal. I knew that a single reservation would have been catastrophic. As each member agreed, my suspense did not lessen, because Appleton was on the chairman's right and the comments and approvals were moving around from the left. At last we reached him. He could have said simply, "I agree." He did not do that, and in later years I was to be given frequent reminders of his prophetic words. "I am," he said "impressed by the wide range of problems in astronomy and geophysics which Professor Lovell has listed as capable of solution by a radio telescope of this size, but I am even more impressed by the possible uses of this instrument in fields of research which we cannot yet envisage."

The minutes record that letters of approval were read from Massey and Ratcliffe and that "it was agreed that Professor Blackett and Dr. Lovell should forthwith make application to DSIR for a grant in aid, together with an immediate allocation to cover the cost of exploratory work . . . it was further agreed to invite the Council of the Society to support in the strongest terms this project which the Committee believed would be of vital importance to the future development of astronomy in Britain."

At the time of the Edinburgh meeting the minutes record that "the cost of erection and provision of ancillary equipment would lie between £50,000 and £100,000." This was the range of estimates I had already been given by Sisson of Grubb Parsons and by Roberts of Coubro & Scrutton during my earlier contacts, already described. The meeting with Husband on 14 June was the first real business contact, and although, as he confirmed in a letter the next day, he thought we would be "reasonably safe" with a figure of £100,000, he emphasized that for such an unusual project it would be essential to carry out a detailed design study in order to obtain firm tenders. In fact, we entered a figure of £120,000 in the application to the DSIR to include the additional cost of a control building and other miscellaneous items. These items included Husband's fee for the design study of £3,000 and £300 to sink a borehole on the proposed site in order to ascertain the foundation structure that would be required.

When I reflect on the long series of committees through which such a request would be routed today, I am astonished that just over a week later, on 22 June, I was summoned to a meeting of the Research Grants Committee of the DSIR and once more presented my case for the telescope. There was not the slightest problem. The committee included P. I. Dee and other colleagues whom I knew and had worked with during the war, and they seemed as eager to help me as I was to get on with the telescope. I came out of the meeting into a sunny London street with immediate authority to proceed. At that time the offices of the DSIR were near Trafalgar Square, from where one has a fine view of Nelson's column poised among the fountains and pigeons in the center of the square. The column holds the statue 180 feet above the London streets. I turned to gaze at it and realized with a tremendous thrill that this was merely the height of the trunnion axis of the telescope I was proposing to build. That was to be one of the few moments of happiness for many, many years in my efforts to make that dream a reality. In retrospect two circumstances about that day strike me as extraordinary. One is that, unknown to Husband, I was still in the thick of the discussions with Head Wrightson (the calamitous failure of contact did not occur until 27 June, and it was 18 July before I wrote to make a final termination of our arrangement for the firm to design the telescope). The second concerns the figure of £100,000, with which Husband thought we would be "reasonably safe." Only two months before Husband gave me this figure, we had been told by Ryan of Head Wrightson that a *200*-foot paraboloid would cost at least £150,000. It was never my job during the war to worry about the cost of projects. The major concern was whether the job could be done and getting it done as soon as possible. So it was now; my thoughts were entirely concentrated on the fact that we were about to design a huge telescope that could and would be built.

13

The First Design Study,
June 1950–March 1951

At the moment on 22 June 1950 when I emerged so joyfully from the DSIR meeting with authority to proceed with the design study, I had absolutely no idea of the task that lay ahead. During that summer and autumn I began to feel that it was not like designing and producing complex equipment for a bomber aircraft; rather it was as though one had to design and build the bomber aircraft as well. Immediately my whole being became immersed in the seemingly endless details not only of liaison with Husband but of working out with my colleagues—mainly Hanbury Brown, Clegg, and Davies—the minutiae of what we ourselves wanted. All the time Patrick Blackett was hovering and frequently descending like an eagle to tear our proposals to pieces.

Husband had mentioned a time of about six months to complete the design study in enough detail to enable us to return to the DSIR with a firm price. There was much detail to be dealt with, and on 5 July he came to Jodrell for our first design meeting. I could not know that for many years after that meeting it would be rare for a day to pass without some form of contact between Husband and myself. By telephone or letter the thread of communication never ceased. It is ironic that years later a government committee pilloried us for lack of contact. Rows of filing cabinets, the contents of which now refresh my memory of those years, are the lasting evidence of a collaboration that continued for three decades—and not always without dispute or controversy. Husband's office was in Sheffield—a mere forty miles from Jodrell, although the intervening Pennine hills provided a physical barrier in the winter months that often stopped me from traveling to Sheffield but rarely prevented a more resolute Husband from making the westward journey to Jodrell.

At that first meeting, on 5 July 1950, I had Hanbury Brown and Clegg with me. Blackett came for only a part of the meeting, during which, quite typically, he made a suggestion that eventually determined how we would drive the telescope. Husband was primarily a structural engineer, and at that time he had no expert mechanical engineers on his staff. Blackett, remembering his naval career, said that the problem was similar to that of aiming and controlling the gun of a battleship. He told me to visit the Admiralty Gunnery Establishment at Teddington, in Middlesex, and so I arranged with J. M. Ford to be there on 20 July with Husband, Clegg, and J. G. Davies. Ford and his naval colleagues thought our problem was simple in comparison with that of controlling the gun turret of a battleship and advised us to talk to Metropolitan Vickers about the use of their "metadyne" control system. This was an electrical drive system that would give the telescope an acceleration proportional to the displacement from the required position. By this time Husband had become concerned about the size of the gear racks that would be needed to drive the elevation. Ford also solved that problem. He advised Husband to obtain some of the 25-foot diameter racks used to drive the 15-inch gun turrets in a battleship.

Husband soon secured a major bargain over the question of these large racks. On 3 August he journeyed to the ship-breaker's yards in Inverkeithing, in Scotland, where the battleships *Royal Sovereign* and *Revenge* were being broken up, and discovered that one of the *Royal Sovereign*'s turrets had not yet been broken down and that two complete 15-inch gun turret racks of the *Revenge* were in perfect condition. The problem was to arrange with the ship-breakers to salvage these racks in good condition. Although we were very far from any positive decision to build the telescope, the question of securing these racks was urgent, since the scrap value would be only a few pounds per ton, but, as Husband warned me a few days after his visit, "to produce two new racks and pinions of this quality would cost several thousand pounds." By mid-September, Thos. W. Ward, the ship-breakers, told Husband that the rack and pinion from the *Royal Sovereign* was freed and lying aboard the ship and that we could have it for £250 and also the two from the *Revenge* when they had been freed. By any standards this was a bargain offer, but we had the problem of where to find the £1,000 to cover their cost and transport. The DSIR refused to help: fortunately the university allowed me to meet the cost from the Jodrell allowance, and, remarkable though it may seem today, that was a major sacrifice from our research budget. Long before we had any authority to build the telescope, these racks and pinions arrived at Jodrell—and they lay in and near our workshop for many years as a symbol of hope that was to be long in realization.

All the large optical telescopes so far constructed had been on an equatorial mount—that is, their supporting column was built at an angle to the horizontal so that the main axis of the telescope was parallel to the earth's axis of rotation. We were proposing to build an instrument with a 250-foot-diameter bowl, and it was considered to be out of the question to construct this on an inclined axis. Our mounting was to be alt-azimuth—that is, the main framework of the telescope would rotate on a railway track (motion in azimuth), and the bowl, suspended on an axis over 180 feet above the ground, could be tilted (elevation or altitude motion) independently of the rotation. Although this would make it easy to drive the telescope to a given azimuth and elevation, we also wanted to be able to follow a star or other object in the heavens. In other words, although the fundamental motions of the telescope were in altitude and azimuth, we had to devise a converter system so that the rates of movement in azimuth and elevation would produce a motion so that the narrow beam of the telescope could follow precisely an object in the heavens: that is, a sidereal motion.

J. G. Davies was given the task of designing this device. The essence of the problem is to solve the fundamental equations of spherical trigonometry. Today, with digital computers, this would be an elementary problem. In 1950, when there were no such computers, an analogue solution had to be sought. Davies developed the equations that had to be solved and designed the circuits to include magslip resolvers. We had used these devices in some of our wartime radar equipment. Essentially they consisted of a rotor, into which 50-cycle alternating current was fed, rotating around a stator with special windings. There were several types of magslips. The ones required by Davies gave a voltage output proportional to the sine and cosine of the rotor's angle of rotation. Eventually the control system for the telescope was to use nearly fifty magslips of various types, but in the autumn of 1950 we urgently needed five so that Davies could demonstrate that his proposals would work, and without the money to purchase these we had the greatest difficulty in borrowing them. Eventually, as the end of 1950 approached, Davies triumphantly produced a board of gadgets with the five magslips arranged to solve the appropriate equations and give voltage outputs that could be applied to the azimuth and elevation driving systems. This was the essential core of the complicated system that was eventually transformed into the control system of the telescope to drive it in sidereal motion.

At our early summer meetings with Husband it was agreed that we would have to initiate action on the control and driving system. Our misjudgment about the magnitude of the project we were undertaking is indicated by the fact that our conversations with Sisson assumed that we would be concerned with a telescope weight of about two hundred tons (which ultimately turned out to be an underestimate of nearly ten times). Although very helpful, Sisson was extraordinarily gloomy about the likelihood of his firm's undertaking any of the work. Surprisingly he was fighting for survival because the optical works section of the Grubb Parsons organization was a severe financial drain on the company.* The only hope he gave us was that his firm might make the control panel. Early in 1951 he wrote a tedious letter to Husband about the difficulty and costs of dealing with government departments and advised us to include a figure of £20,000 for the

---

*The optical works of Grubb Parsons closed finally in the early 1980s.

control panel alone. This was five times the figure we had in mind and effectively terminated our connection with Sisson and Grubb Parsons.

Sisson was my initial outside contact in connection with the telescope proposal. Although our dealings with him and the firm with which he put us in touch came to nothing, he was the first industrial engineer to become excited about the problem and the possibility of building such a telescope—and for that early encouragement in 1949 I was always grateful to him.

After the failure of this effort to involve Sisson and his firm, I turned to Metropolitan Vickers in Trafford Park, Manchester. Fortunately I had a long history of contact with that firm, and on 26 July I went with Davies to Trafford Park. A brief note I made of that meeting simply recorded that the firm thought the problem "was a fairly simple one and the manufacture of a suitable metadyne driving system would present no difficulties." However, what should have been a simple arrangement made with colleagues and friends in Metropolitan Vickers eventually turned into one of our major nightmares, and, as will be seen, many years later we were forced to abandon our contacts with that firm under acrimonious circumstances.

Meanwhile, throughout these summer and autumn months of 1950, Husband proceeded with the structural design and calculations. The time would eventually come when computer programs would be available for the rapid evaluation of the stresses and strains in a structure of this magnitude. In 1950 no such possibility existed. Only a slide rule and a hand calculating machine graced Husband's large desk in the Sheffield offices, and soon we realized that his estimate of six months for the design was impossible of fulfillment. At that time we were intending to make the reflecting surface of 2-by-2-inch wire mesh, and there seemed to be no available data on the wind forces at various speeds and inclinations and under icing conditions. I learned that it was an extraordinarily difficult matter to calculate the forces on a huge airfoil of the paraboloidal shape we were proposing. First, I appealed to W. A. Mair in the university, who had developed a "fluid motion" laboratory on a small private airfield north of Manchester. Husband got some data from him by the middle of

FIGURE 1: The staff of the H. H. Wills Physics Laboratory, University of Bristol, 1935. The author is in the back row, second from the right. In front of him, near his right, stands C. F. Powell, with the glassblower J. H. Burrow on Powell's left. A. M. Tyndall is center front. On his right is N. F. Mott and on his left (arms folded) S. H. Piper. The author's supervisor, E. T. S. Appleyard (wearing spectacles), is in the second row, fifth from the left. Klaus Fuchs, who was imprisoned after the war as a traitor for revealing atomic secrets to the Russians, is in the second row, third from the left. The distinguished theoretical physicist W. Heitler is seated in the front row on the extreme right. H. W. B. Skinner is also seated, second from the left.

FIGURE 2: The first Halifax bomber V-9977 to be equipped with H₂S. The photograph was taken shortly after the bomber arrived at Hurn airport on 27 March 1942. The perspex cupola housing the H₂S radar scanner can be seen immediately beneath the Royal Air Force insignia. This aircraft, the crew and research staff, and the only experimental H₂S equipment were destroyed when it crashed in flames on 7 June 1942.

FIGURE 3: The 3-centimeter H$_2$S picture of Berlin as seen on the cathode-ray tube of a Lancaster bomber of the Pathfinder Force during the run up to bomb release during the night of 1–2 January 1944. The bomber was at 19,000 feet. The straight line running nearly horizontally to the right of the picture is the heading of the aircraft. The black circular area is immediately beneath the bomber; beyond, the line of the Tegeler See and Wannsee and other features of the city are visible.

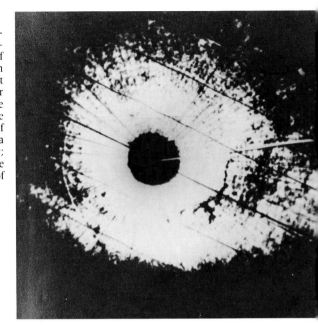

FIGURE 4: December 1945, the first day at Jodrell Bank. The receiver trailer of the army radar equipment near the wooden huts used by the university botany department.

FIGURE 5: A detail of the 218-foot transit telescope in 1947 showing a section of the completed paraboloidal reflector of 16-gauge galvanized steel wire.

FIGURE 6: A section of the staff photograph of the physics laboratories of Manchester in 1949. P. M. S. Blackett is in the center, the author is on his left, and L. Rosenfeld is on Blackett's right.

FIGURE 7: The staff of Jodrell Bank in 1951, in front of the searchlight aerial. J. A. Clegg stands on the author's left and J. G. Davies on his right. On his right in the front row stand J. S. Greenhow, C. Hazard, and Mary Almond.

FIGURE 8: The trunnion assembly being hoisted to the top of one of the towers on 14 September 1955. The large spherical bearing is to the right, and one of the gun turret racks (figure 9) was fixed to the framework on the left of the assembly on 30 September 1955.

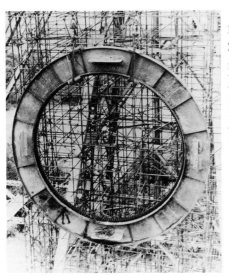

FIGURE 9: The bowl is driven in elevation through racks and pinions mounted on the tops of the two towers. Two pinions, each driven by 50-horsepower electric motors through a 1-to-22,240 gear reduction, mesh with each rack. The racks are fixed to the bowl framework through the 30-ton girder assembly visible in figure 10. The racks, 27 feet in diameter, are from the 15-inch gun turrets of the dismantled battleships HMS *Royal Sovereign* and HMS *Revenge*. Here, on 30 September 1955, one of these gun turret racks is being hoisted to be fixed to the trunnion assembly visible in figure 8.

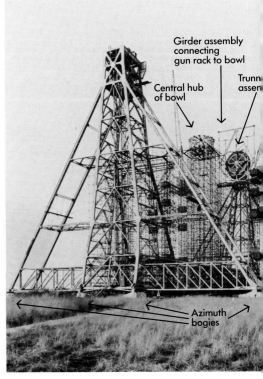

FIGURE 10 RIGHT: The year 1955 marked a vital stage in the building of the telescope: the structure as it appeared in October. On 14 September the trunnion assembly (figure 8) was lifted to the top of the farther tower. One of the gun racks (figure 9) through which the telescope is driven in elevation was fixed to the assembly on 30 September. The 30-ton girder assembly—the meeting point of the gun rack and the bowl—was fixed in position on 2 October. Three weeks later the central hub of the bowl framework was hoisted to the top of the supporting scaffolding tower.

Stabilizing girder (the "bicycle wheel")

Housing of trunnion and elevation driving system.

Azimuth railway track

Azimuth bogies

FIGURE 11: The completed Mark I radio telescope photographed in the early 1960s.

FIGURE 13: The cartoon by Papas published by the *Guardian* on 3 March 1960. The author is holding the telescope as a begging bowl behind the three armed services, whose estimates included a multimillion-pound sum for the construction of the Fylingdales defense radar. (Reprinted courtesy of the *Guardian*.)

FIGURE 12: The author and his wife Joyce with their children at The Quinta, Swettenham, summer mid-1950s.

FIGURE 14: The Solvay conference in Brussels on the theme "The Structure and Evolution of the Universe," photographed on 14 June 1958. The author is standing second from the right, immediately behind Harlow Shapley. Many other astronomers and scientists mentioned in this book are in the group. J. R. Oppenheimer is seated fourth from right. To his left is W. L. Bragg, president of the conference. On Bragg's right is W. Pauli. J. H. Oort is seated second from left, next to the Abbé Lemaître.

FIGURE 15: The author with his wife at Buckingham Palace in February 1961, after receiving the accolade of knighthood at an investiture held by Her Majesty Queen Elizabeth the Queen Mother. (Reprinted courtesy of the Press Association.)

FIGURE 16: Her Royal Highness Princess Margaret pictured on the radio telescope with H. C. Husband and the author, 18 March 1960. (Reprinted courtesy of the *Congleton Chronicle*.)

FIGURE 17: The academician M. V. Keldysh, president of the Soviet Academy of Sciences, with the author at Jodrell Bank, February 1965. (Reprinted courtesy of the *Daily Telegraph.*)

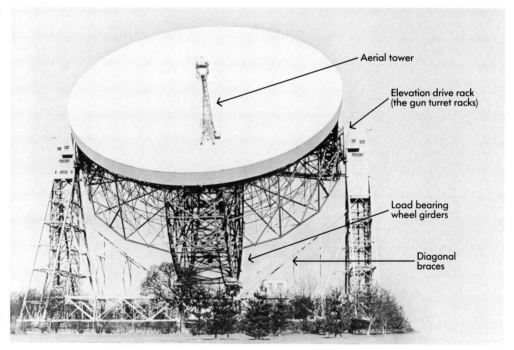

Aerial tower

Elevation drive rack
(the gun turret racks)

Load bearing
wheel girders

Diagonal
braces

FIGURE 18: The Mark IA telescope. The major change compared with the original Mark I (figure 11) is the new reflector mounted above the original, which is still in place. The aerial tower has been lengthened and strengthened to carry a heavier load of equipment near the focus. The slender stabilizing circular girder (the "bicycle wheel") seen in figure 11 has been replaced by the massive double load-bearing wheel girders. The diagonal braces from the towers to the diametral girder that were introduced after the hurricane of 2 January 1976 are clearly visible.

FIGURE 19: The author in conversation with Margaret Thatcher, then secretary of state for education and science, during her visit to Jodrell Bank on 5 May 1975. (Reprinted courtesy of the *Guardian*.)

FIGURE 20: The author with his wife in the gardens of their home, The Quinta, in Swettenham, early March 1989. (Reprinted courtesy of the *Manchester Evening News*.)

September and then wanted more tests carried out at the National Physical Laboratory. It was already December before Husband had these relevant data for the vital calculations on which the operation and safety of the telescope would depend. In the event, it was 20 March 1951 before we could complete our formal application—a delay of a few months that then worried me very much but that I soon learned was of little consequence in the time scales that were to develop.

Early in 1951, impatient at the delay in completing the design and when we were beginning to get estimates from Husband about the effect of wind on the structure, Blackett commenced checking various points himself from a fundamental physical approach, at which he was masterly. He made me produce a long memorandum to justify our calculations of the power of the motors required to drive the telescope in various wind speeds and then concluded that our proposals for the metadyne system would produce a jerky movement. That meant more calculations and a memorandum dealing with factors about the effect of wind and friction on the motion of the telescope—factors only dimly apprehended at that stage.

Toward the end of January 1951 we had managed to reassure Blackett on these and other fundamental engineering and scientific issues. I had become utterly weary of spending Saturdays and Sundays—the days Blackett seemed to reserve for our affairs—on these matters when my spirit wanted to be in church or in the garden. He then casually remarked, "Of course, you will produce a comprehensive report on all these matters to accompany the formal application to the DSIR." I had already written large numbers of reports and memoranda and saw no reason for producing more, but Blackett was adamant and said he had merely been searching for the justification that members of the committee would seek before allowing me to embark on this huge project.

This report became the "blue book," the cornerstone and scientific case of our plans for the telescope. We had to make seventy copies (because the DSIR wanted sixty). That would not be difficult with today's copying and binding techniques, but in 1950 we had to make a wax stencil of each page and turn the handle of the duplicating machine seventy times for each of the eighty pages. I had sought the help of a friend in the publishing business, but the time scale and cost made this approach out of the question. A local shop bound the papers in blue cloth, and when we finally dispatched our application to the DSIR, on 20 March 1951, one blue book was included. Sixty more went by passenger train a week later. No doubt most of these blue books have long since been shredded, but at least one has survived. It is on my desk now as I write this thirty-six years later. A poor thing by the standards of today's expensive and glossy brochures that accompany even minor applications for money, but for me a treasured item.

This blue book and a memorandum from Husband accompanied the formal application to the DSIR, which we signed, at last, on 20 March 1951. Late in January before the submission, however, I went to Sheffield to discuss with Husband the preparation of his memorandum. On my return I wrote to Blackett that I had come away "rather depressed about the latest financial situation with regard to the paraboloid." "Costs have risen so steeply," I said, "that it is now quite certain that our figure of £120,000 . . . is too low . . . and it begins to look as if the cost of the completed structure is unlikely to be less than £150,000." A month later Husband produced his memorandum with the even larger estimate of £242,000, that is, twice the figure we had entered in our application a year earlier. It was the beginning of a story that was to become depressingly familiar, with none of the parties to the project being blameless. Although I had placed the blame on steeply rising costs in my warning note to Blackett at the end of January, that was a small part of the story—less than £20,000. Two major disparities arose because Husband had underestimated the costs of the foundations by £20,000 and the driving system by a similar amount. A third was our concern. We had originally intended the reflecting mesh to be about the same as that of the 218-foot transit telescope and had specified an open-mesh reflector of 4-by-4-inch section. When we had authority to go ahead with the design, in June 1950, we proceeded to study the radio frequency requirements in more detail and for the first time calculated the leakage of the radiation through a mesh of this size. To our alarm we discovered that at a wavelength of 1 meter (the shortest at which we were proposing to work) some 20 percent of the incoming radiation would be lost through the mesh, whereas we would lose only about 2 percent with a 2-inch mesh. Husband estimated that this had added another £20,000 to the estimate because of the need to increase the amount of steel in the structure to take account of the increased wind loading.

When Blackett saw these figures, he was furious and once again insisted on checking all the calculations and the breakdown of the detailed costs on every single item. He complained bitterly about my obtuseness in insisting on such a large paraboloid and said the chances of getting nearly £250,000 were quite negligible, whereas with a smaller telescope we could still have done a great deal for British astronomy for something in the neighborhood of £100,000. For my part I was thankful that, at last, the design was complete, but it was a thankfulness replaced by a deepening anxiety, as month by month we awaited some indication of the DSIR's reaction to this request for £250,000.

14

The Building of the Telescope: The Year of Decision

My request to the DSIR for £259,000 was more than double the sum I had suggested to the Research Grants Committee in June 1950, when they had so readily agreed to make the grant to carry out the design study. By the standards of the 1980s the amounts of money we were seeking in 1951 may seem small. Inflation and the growth of the science budget have eroded the significance of a sum of a quarter of a million pounds in 1951. I knew from Blackett and others that because of the country's financial crisis the DSIR had been forced to suspend all but the essential and minor applications for research grants. Evidently the only means open to the DSIR was to seek another special grant from the Treasury for the telescope. In time I understood why my personal contact with the Research Grants Committee—M. A. Vernon—had informed me on 17 May that new circumstances made it necessary to postpone a scheduled meeting of Husband and myself with the committee.

I believe there are two reasons why, thirty-five years ago, the DSIR merely postponed a decision instead of rejecting my application. The first is that in those days, only six years after the war, science and scientists were held in the highest regard. The scientific contributions to the Allied victory had not been forgotten, and faith in the scientists and their work had not then begun its steep decline. The second reason is more personal. The grant-giving body of the DSIR was composed of scientists with whom I had been associated in many wartime activities, and I was too young for any personal animosities or ambitions to have developed. In fact, P. I. Dee, my senior associate during the wartime years, was the chairman of the Research Grants Committee. On the one hand this desire to help prevailed, and on the other powerful external influences were active on my behalf. These influences were centered around four individuals—Blackett, Sir Henry Tizard, Lord (Ernest) Simon, and Sir John Stopford.

The part that Blackett had played in my scientific career and his enthusiasm for the telescope will have been evident from the earlier parts of this book. His reputation as a scientist was so high that no one in the country would lightly disregard his opinion. At that time in 1951 he was emerging from the nadir of his immediate postwar political fortunes. His important relationship with Harold Wilson (the future Labour prime minister) had begun in 1949 when Wilson, as president of the Board of Trade, had appointed him to the National Research and Development Corporation. A year later Blackett became the senior member of a small group of eminent scientists who held regular meetings with the leaders of the Labour party (first Hugh Gaitskell and then Harold Wilson) in order to evolve a scientific and technological policy for the country. This eventually led to the radical changes in the country's scientific organization and policy that occurred when Wilson became prime minister, after the general election of October 1964, but in 1951 it was Blackett's close contact with other members of that group that was so important to my own affairs. In particular Sir Ben Lockspeiser, the secretary of the DSIR, was a member of this group. Lockspeiser's kindly disposition toward me had this origin, and the dire events that beset me when he retired from the DSIR in 1956 will appear later in this story.

I had met Tizard briefly before the war when he visited the physics department of the University of Bristol. The circumstances of this meeting were entirely accidental. Tizard, then rector of Imperial College, London, had such a high regard for the city of Bristol that he arranged for one of his sons to spend some of his student days in the university. At that time I was a postgraduate student doing occasional duty as a demonstrator in the practical laboratory class. One afternoon Tyndall brought Tizard into this class to see his son, to whom I was attempting to explain the complexities of the compound pendulum. Only after the war did I realize that in those immediate prewar years he was a member of the Committee for the Scientific Study of Air Defense, and chairman of the famous "Tizard committee," working in great secrecy to foster the development of radar. Neither did I know that Blackett was one of the key members of that small committee. That tangled skein of circumstances may well explain why Tizard always sought me during his many visits to TRE during the war to ask how I was faring and to discuss the radar developments on which I was working.

The antagonisms between the Tizard-Blackett group and Lord Cherwell in the late 1930s and during the wartime years were subsequently well publicized. However, in the general election of 1946 the defeat of Prime Minister Churchill eroded Cherwell's influence, and in the Labour government of Clement Attlee, Tizard became a person of great power and influence. I had not seen him since the end of the war, but in early July 1951 another accidental circumstance made me aware that he was deeply involved in the attempts to persuade the Treasury to finance my telescope project. The occasion was the opening by Tizard of new research laboratories in Manchester for Simon Carves Ltd. I had been invited to the ceremony and to the luncheon on 3 July, and Blackett had asked me to look after Tizard in the afternoon because he wanted to see Jodrell. Would I then drive him to Crewe so that he could get an evening train to London? Tizard was old enough to be a father figure to me, and during those journeys he seemed to understand what effect the rejection of the telescope project would have had on me. He warned me in the kindest possible manner to be prepared for failure, because there were acute financial difficulties. As I waited on the platform with him for the London train, he promised to do everything possible inside the Treasury. Thirty-five years later the memory of that afternoon is vivid. I knew then that powerful influences were trying to force the Treasury to make an exception to the moratorium on the finance of major new scientific projects.

Whether this Tizard-Blackett axis would have succeeded without the offer of help from another source I do not know, but I realize now that two other key figures were supporting my cause at the highest levels. Two remarkable men reigned supreme in the University of Manchester. Lord Simon, the chairman of the university council, was a wealthy industrialist. Having successively been chairman and governing director, he was now president of the Henry Simon engineering enterprise and was at that time the governing director of Simon Carves, whose new research laboratories Tizard had opened on 3 July. In addition to these engineering enterprises, Ernest Simon was a major figure in many diverse political and social fields. In the 1920s he had been Lord Mayor of Manchester and a member of Parliament as a Liberal, but in 1946 he changed his allegiance to the Labour party. In this postwar period, as a Labour peer he was chairman of the British Broadcasting Corporation. During the wartime administration of the coalition government he had been the deputy chairman of the Central Council for Works and Buildings. This background made it understandable that Lord Simon was most happy when involved with new ideas and large projects. In 1949, when we were still working in a muddy field, he asked the vice-chancellor to bring him to Jodrell. He came with the chancellor, Lord (Frederick) Woolton, and from that moment he spared no effort to help me. On the day I received Vernon's letter of 17 May postponing the meeting with the DSIR, I was in the university. While walking across the university quadrangle, I encountered Lord Simon, who was so dismayed at the news of delay in what he regarded as a small financial matter that he instantly spoke to Gaitskell, then leader of the Labour party, and Addison (Viscount Addison, the lord president of the council). This was typical of Lord Simon's impatient enthusiasm for the telescope project, and he never once wavered in his support throughout all the uncertain years that lay ahead.

The last of this quartet, Sir John Stopford, the vice-chancellor of the university, I have already mentioned in earlier chapters. That he was a distinguished scientist and a fellow of the Royal Society is perhaps one of the most significant factors in this whole complex story. However, that alone would not have given him the faith and power to help. It was all very well for Blackett to espouse my cause before him, but Stopford, a wise man of long experience in university affairs, had seen too many of his staff forsake their posts for domestic and other reasons. Of course, he judged a man by his academic standing, but he sought a deeper insight into the stability of the individual by scrutinizing the background. In later years, when our relationship had become less formal, he said, and not in jest, that in assessing a man's suitability for the academic career he would talk to the man's wife and discover a little of the family background. That may seem old-fashioned today, but in my case he discovered a family fortress that, he rightly judged, would never yield in times of trouble. With his faith established, he proceeded to exercise his power. His major extramural activities lay in the medical field, but he held one office of crucial importance as far as the telescope project was concerned. He was the long-serving and influential vice-chairman of the Nuffield Foundation, and it was through that avenue and the interlocking influence of Blackett, Lockspeiser, Tizard, and Simon that the solution to the immediate monetary problem was found.

I do not know at what stage the sharing of the burden of the cost emerged as a possible solution, but thirty-five years later the deputy director of the Nuffield Foundation kindly responded to my request to search the foundation's archives of 1951. On 10 July of that year—that is, one week after his visit to Jodrell—Tizard wrote to its secretary, assuring him that he had not told me, Blackett, or Lockspeiser that he would make an approach to the foundation. Tizard had suggested to its trustees that they should bear the whole cost of the telescope. Lord Nuffield was consulted and gave his consent, but the trustees decided to wait and see how much money the DSIR would offer. During the summer and winter of 1951 the problem remained with the DSIR. At the time of Tizard's visit in early July the project had been approved unanimously by the Research Grants Committee of the DSIR and was then before the DSIR Advisory Council.

I knew that by the end of July the project had been thoroughly approved by all the DSIR committees and by an external advisory group Lockspeiser had convened. With this backing Lockspeiser approached the Lord President of the council and the Treasury for an additional grant so that the DSIR could meet the cost. They, too, approved of the project but would not commit government finances to increase the DSIR allocation, because of the political uncertainties. These uncertainties were resolved in the autumn general election of 1951, but not in any way that gave immediate assistance to the DSIR for the financing of the telescope. The Labour government was defeated and the Conservatives returned to power, with Churchill once more prime minister. The proposal for the telescope was not, in itself, a political issue, but Churchill again appointed Lord Cherwell to high office as paymaster general, with the consequence that the power of the Tizard-Blackett grouping instantly waned.

During the war I was too young and too occupied with the radar projects that were my responsibility to understand or take sides in the Tizard-Cherwell antagonisms. In fact, it was Cherwell who had forced through the highest priority for my H_2S bombing aid, and he often came to TRE for a flight in the bomber aircraft in which we had installed experimental equipment. I got on as well with Cherwell as I did with Tizard; in addition, I had no strong political affiliations, so in the autumn of 1951 I sought help from this new power grouping. First, I approached Cherwell but without the hoped-for success. On 8 November he wrote to say he agreed about the telescope but did not know enough about the technical details and suggested that we might save money by making it smaller. That was the very last thing I contemplated and so abandoned that avenue in favor of an approach through the vice-chancellor to Lord Woolton. On paper, at least, that avenue should have been excellent. Lord Woolton was the chancellor of the university; moreover, in this new Churchill administration he had been appointed Lord President of the council, with direct responsibility for the DSIR. I soon discovered that my contacts in the DSIR regarded Woolton's interest as a very favorable sign and that they expected a decision "one way or the other" by the end of January.

With the return to power of the Conservatives another important pressure point fortuitously opened in November. In the chapters about the war I wrote of Sir Robert Renwick and of his
emissary Frank Sayers. During the war Renwick was a person of
immense influence. He enjoyed direct access to Prime Minister
Churchill and had been appointed by him to cooperate with Dee
and myself so that every possible priority was given to the development and installation of the H$_2$S blind-bombing equipment.
Scarcely a day would pass without a telephone or personal contact between Renwick and myself, and Frank Sayers was with us
constantly. At the end of the war these contacts ceased. Renwick
returned to his business interests and Sayers into industry. Now,
at this critical moment in November 1951, the avenue reopened.
Sayers had read about our work at Jodrell and asked if he could
visit me. He did so in the third week of November, at the moment
when I was full of gloom about the prospects for the telescope.
His approach now, as it always had been during the war in times
of trouble, was to seek the help of Renwick. Sayers asked for a
full brief, and on 27 November I sent him a long account of the
project and of the efforts to finance it through the DSIR. I was
never able to discover what happened, although if wartime prec-
edents were a guide I have no doubt that Renwick telephoned the
prime minister—as he did in my presence on another occasion
more than ten years later. After the final resolution of the problem, I heard from Sayers, "If you have not already done so, I
think you should give some thanks to Sir Robert because I can
assure you that we have worried the life out of various people."

At the beginning of 1952, however, the possibility of financing
the telescope remained delicately balanced. In February I was
asked to bring the cost figures in the application of March 1951
up to date on the assumption that the program would start on 1
April 1952. In collaboration with Husband, I revised the application and estimated the annual expenditure on the basis that the
construction would take three and a half years—a year longer
than we had originally stated, because of increasing delays in the
supplies of steel and other material. The 1951 application was for
£259,000; increased costs had now raised the figure to £335,000.
There followed another four weeks of silence and suspense until,

one afternoon late in March, Blackett telephoned from London with the simple but pregnant message "You're through."

I received the official news by phone, and a few days later a letter confirmed that the DSIR had now been informed by the Nuffield Foundation that it was prepared to share equally with the DSIR the cost of the radio telescope project. There were two conditions: one concerned a board of visitors to settle any disputes about the program when the telescope came into operation; the other required that I "would make a point of seeing the project through and remain at Manchester to be in charge of the project for not less than five years." It was Good Friday when I read this letter, and as often on that day I was listening to *Parsifal*. Klingsor had been vanquished, but I had no vision that so much magic would come into my life or that I would so narrowly escape the fate of Amfortas.

15

The Building of the Telescope: The Early Years of Construction, 1952–1953

When I was a young man, I was inclined to keep a diary. I never had the self-discipline to do this with great regularity, and when I look at the books of writing now I find that the entries are concerned mostly with the cricket matches in which I played and with music. In the early 1930s, for example, I can find regular details of my batting average or bowling performances, and without too much difficulty I could discover how many times during the year I had listened to a Bach fugue or a Beethoven symphony. I could also provide a fairly accurate account of the weather wherever I happened to be, simply because that affected the cricket.

After I married, in 1937, the entries became more sporadic but also more serious, particularly as the war began to cast its shadow. I continued the diary at irregular and increasingly

220

lengthy intervals during the war. I am glad that I did so and that those personal wartime diaries are intact because they have helped settle many arguments about the dates and sequence of the radar developments in which I was involved. At the end of the war I wrote no more until the day in April 1952 when I received the letter from the DSIR announcing the grant for the telescope. For some reason I made a brief diary note of this event. I had a peculiar feeling that it would be important to keep a record of the building of the telescope, and I did so. It is on my desk now as I write, four hundred pages of typed foolscap sheets. It is a record not merely of the day-to-day events on the site but also of all my dealings with Husband and with anyone else concerned with the telescope. In the moments of great crisis that arose I told Rainford that I intended to burn the diary. He stopped me from doing so: "It may be the only record that saves you." Indeed, the diary entries of my constant contact with Husband eventually forced a unique withdrawal of evidence that had been given before the Public Accounts Committee and an apology from that committee which made it possible for me and the telescope to survive. These four hundred pages, as well as the dozens of files of letters and reports, also enabled me to write the detailed story of the building of the telescope in those years from 1952 to 1957. *The Story of Jodrell Bank,** published in 1968, is that record, and no point would be served now in once more reproducing that full history. In these two chapters I shall summarize the timetable of construction, but more especially the reasons why such a succession of crises developed over those five years—it was five years, not the optimistic two and a half years of our 1951 application or the revised three years of the 1952 estimate at the time of the award of the grant.

More than thirty-five years later it may well be hard to understand why there was so much difficulty and delay over what now appears to be a rather small sum of money. During the five years of construction the cost of the telescope doubled. This escalation was due not only to rising prices but also to the changing decisions we were forced to make for both engineering and scientific

---

*Published by Oxford University Press and by Harper & Row in 1968; republished as *Voice of the Universe*, with additional chapters, by Praeger in 1987.

reasons after the construction commenced. The building of the telescope was a unique engineering and scientific adventure. Furthermore, it was one of the first ventures in building a large-scale facility for research. Neither the university nor the government scientific-grant-giving bodies were yet accustomed to massive scientific enterprises, and these various factors caused tremendous pressures to be exerted on the project.

The cost that the DSIR and the Nuffield Foundation agreed to share in March 1952 was £335,000. The complexities during the five years of the building of the telescope that led to a more than doubling of this cost have this background, and these developments can be understood only against the relative magnitude and cost of the telescope as it might appear in the world of today. Today (1986) by a straightforward adjustment of the UK government's cost-of-living index, this figure would have to be increased twelvefold—say, to £4 million. However, no one believes that a telescope of this size could be built for that amount. Because the costs of such specialized construction work have risen much more steeply than the conventional price index, the estimated cost of constructing the telescope today would be nearer £20 million.

There is another important factor. In 1952 the grant of £335,000 for the telescope was equivalent to the amount of money at the disposal of the DSIR in any one year for the support of all fundamental research. If I applied today to the DSIR's successor—the Science and Engineering Research Council (SERC)—for a grant of £20 million to build a telescope, that would represent only about a fifteenth of the amount of money the council disposes for research every year. Furthermore, the council would not entertain a grant of this magnitude for an individual enterprise; it would be considered only in the context of a European or wider international enterprise. The expenditure from any such grant would be controlled and scrutinized by a series of committees. In 1952 the grant for the telescope was made to the University of Manchester for my use, and no such safeguards were enforced. If they had been, I do not believe the telescope would have been built, and the story I now relate in this book would not have occurred.

The Foundations

The general design of the telescope envisaged two 186-foot-high towers, which would carry the trunnion bearings and driving racks for the elevation motion of the 250-foot diameter paraboloid. The azimuth motion was to be obtained by mounting these two towers on twelve bogies, six under each tower, running on a double railway track 320 feet in diameter. Therefore, the first constructional phase was to build a concrete ring on which the railway track would be mounted. The track had to carry more than a thousand tons of steel and remain level to a fraction of an inch. In preparation for the original 1951 estimates Husband had sampled the substratum of the field in which we intended to build the telescope. Deep cores had revealed that underneath the surface layers of soil there was wet sand and that a secure base of stone and marl (keuper marl) lay about forty feet below ground level.

On this basis, in order to meet the scientific and mechanical requirements that the track should remain level to a sixteenth of an inch, Husband had decided that 160 reinforced concrete piles must be sunk to this keuper marl in order to support the track and the central pivot on which the telescope was to rotate. On top of these piles a 350-foot-diameter circle of reinforced concrete, 20 feet wide and 6 feet deep, would then be built, and on this concrete ring the double railway track would be secured.

This was all standard practice in foundation work, and when we were able to proceed in the spring of 1952 Husband said the construction of the 160 piles would take only two and a half months. We then anticipated that the actual steel structure of the telescope could commence in the spring of 1953. Perhaps this program was far too optimistic, but that it did not have the slightest chance of fulfillment was entirely my fault.

In those days there were still many legacies of the war leading to shortages of material. Government licenses had to be obtained before the reinforcing steel could be ordered, for example. Even with these difficulties tenders had been obtained, the necessary licenses for the steel and building work had been secured, and the piling contractors planned to start work on 1 July. They did not do so, because I wanted to change the site of the telescope by a few hundred yards into a field the university did not own. This site was desirable for scientific reasons concerning the geographical position of the instrument in relation to the transit telescope and for practical reasons of proximity to our existing research buildings, workshops, and powerhouse. If the aged lady who owned this field had lived only for another month or so, the legal tangles of the summer of 1952 in which I found myself enmeshed would not have occurred and the telescope foundations would have been sunk on the original site. However, the lady died at this critical moment, and as soon as we had authority to proceed in the spring of 1952 I informed Husband that the possibility had arisen that the university might be able to purchase the whole farm and so obtain the field in which I wanted the telescope to be built.

I did not know that the ability, and indeed the willingness, of the university to purchase this farm was an entirely different matter from the practicability of doing so. We were soon involved in the intricacies of a family dispute among the five sons who were the beneficiaries of their mother's estate. Naturally Husband constantly pressed me for a decision, and the initial optimism had degenerated to so much pessimism that on the agreed starting date of 1 July the contractors arrived and began boring operations on the original site.

That this field, where the contractors began working, is now a large parking lot for visitors and not the site of the telescope is entirely due to the coincidence that three weeks later the legal tangle was settled in a high-court action. I was lecturing in Leeds when I was handed a telegram to this effect from one of the solicitors involved in the case. If I expected that my instant telephone call to Husband, asking him to transfer the contractors to my preferred site, would lead to similar immediate action, I was to be sadly disillusioned; it was 3 September before the contractor's trucks were able to move into this new site for the telescope.

My desire to transfer the position of the telescope had cost us two valuable months of dry weather. By November the site was a quagmire. There was so much water that a casing for the piles that could not be filled with concrete before nightfall was full of water by daybreak, and the mud-coated workmen would spend several hours pumping water and bucketing out more sand to restore the casing to the state in which they had left it the night before. Progress was not helped by the discovery that the level of the keuper marl was steeply sloping across the site. The sample borings had revealed the keuper marl at a depth of about forty feet, but by mid-December some of the piles had to be sunk to ninety feet.

In May 1953 the last of the 160 piles was finished—the work scheduled for two and a half months had taken nine. By the end of July the massive ring of concrete on the top of the piles had been completed, and by Christmas the tunnel from the central pivot to the site of the control room had been built. The contractors departed, leaving us, after sixteen months' work, with ten thousand tons of reinforced concrete sunk into the ground. At a cost of £50,000 we could look on a 20-foot-wide, 350-foot circle of concrete.

The First Financial Crisis

During 1953 I was far less concerned with this slow progress on the foundations than with many other troubles. When the grant of £335,000 was awarded in the spring of 1952, the formal documents contained a broad breakdown of the estimated cost of the major items according to the data Husband and I supplied to the DSIR. These estimates were considered to be correct, and no one expected deviations of more than about 10 percent. The figure we had given for the foundations was £45,000, and no one was particularly worried about the settlement at £50,000. This was only 10 percent higher than the estimate and was readily understood to have arisen because many of the piles had to be sunk to a greater depth than the trial borings.

The issue of the cost of the steelwork was an entirely different matter. For this we had estimated the figure as £105,000. At an early stage we had accepted Husband's recommendation that the main contract for the supply and erection of the steel superstructure of the telescope should be placed with the United Steel Structural Company Ltd., one of the largest organizations in the country, with the capability of both fabricating the steel and carrying out the erection. The original calculations led to the conclusion that 959 tons of steel would be needed. Our figure of £105,000 was based on this. In the detailing of the steelwork the tonnage had increased to 1,000, and since our estimates of early 1952 had been produced for the DSIR there had been a 17 percent increase in the price of steel. By the end of January, Rainford had concluded that the steelwork would cost £120,000 instead of £105,000, and I was bound to agree with his view. In spite of my arguments he refused to sign the contract until I had explained the increased costs to the DSIR and obtained authority to proceed.

Unfortunately, the case that I had to present to the DSIR on 25 February 1953 was not straightforward. I had to explain that the increase was not a mere £15,000 above the original estimate of £105,000 for the steel but that the review I had recently made with Husband indicated that we required an increase of £92,000 in the grant. I could argue that nearly 20 percent of this arose because of the increase in the price of steel, but the stark facts were that we now required 1,200 tons of steel instead of 1,000; the cost of the central pivot on which the telescope was to rotate was now estimated at £12,000 instead of £5,700; there was a 32 percent increase in the estimate for the bogies and driving system; and we wanted £20,000 for the reflecting mesh instead of the £12,000 estimated.

When I presented these figures to Rainford, he remarked that Husband could not have carried out the design work thoroughly before obtaining the 1952 estimates. It was a theme that recurred with increasing acerbity as the crisis developed. However, that was an unfair accusation. It has never been possible to assign blame for what happened in February 1953. Husband may have been partly responsible, but I had introduced a factor that undoubtedly had a major influence on the rise in costs. Without the slightest expectation that I was about to cause such disruption of the original design, I had asked Husband in the autumn of 1952 if we could reduce the size of the reflecting mesh in the central 100-foot section of the bowl. The reason for my request was an important scientific discovery made almost simultaneously in 1951 by astronomers in Leiden, in Sydney, and at Harvard. The spectral line emission on a wavelength of 21 centimeters had been detected from the gas clouds of neutral hydrogen in the Galaxy. This discovery soon emerged to be of such potential significance to our understanding of the Galaxy that we sought ways and means of making the telescope usable on this shorter wavelength.

On 16 September 1952 I had written to Husband to confirm our many discussions about this problem and asked him to investigate the effect of modifying the mesh in the center of the bowl so that the central portion inside a radius of 50 feet would be formed of mesh 2 inches by 1 inch instead of 2 inches by 2 inches. "Perhaps you would let me know," I added, "if you think that this can be done without altering the steel structure in any way." Husband replied, "The arguments in your letter are quite sound and if the 100-ft aperture with the fine mesh is the absolute limit of your requirements there is no insurmountable problem."

Faced with such a change today, one would adjust a few parameters in the computerized design of the telescope and find out without delay whether any untoward consequences of such a request were likely. In those days there were no computers, and Husband already had five members of his staff with slide rules and hand-operated desk calculating machines working on the problem of detailing the steelwork. In his formal response to my letter he wrote that these structural calculations had already been in progress for several months, that it was tedious work, but that the modifications I now required should not cause "a tremendous amount of extra trouble." It is quite clear on reading the files that I suggested the modification "on the assumption that no major alterations will be required in the steel structure such as would seriously influence the general design or give rise to any appreciable increase in cost."

After my meeting with members of the DSIR secretariat on 25 February it soon became evident that further approaches would have to be made to the Treasury for an additional grant. We were instructed to provide a full explanation of the reasons for this increase in the estimated cost; the vital question, of course, was why we had so badly underestimated the tonnage of steel required. That is where the argument centered, and the extent to which my request for a change in the central mesh influenced the outcome was obscured in a mass of detail and uncertainty. The fact was that no one had ever attempted to build a structure like this and that the effects of wind forces, especially the airfoil effect on the paraboloidal bowl, and the torsional stresses introduced by the variation of wind speed with height above the ground, were at that time not susceptible to accurate calculation. In the event, the analysis presented to the grants committee on 6 May fairly evenly distributed the blame between Husband and myself. For example, in the detailed stress analysis Husband had concluded that the main back girder of the bowl was not stiff enough, and he had decided to add 30 tons of steel at each end of this girder to give more torsional rigidity. The extra wind forces introduced through my request to modify the central section were held to be responsible for another 50 tons of stiffening. Of course, everything interacted. For whatever reason more weight was added in the bowl structure, so the loading on the central pivot and the bogies was increased.

The analysis we eventually produced for a meeting of the DSIR committee on 6 May gave the weight of the bowl structure that would move in elevation as 562 tons and that of the supporting tower structure as 582 tons. There were also miscellaneous items, such as lifts to gain access to the trunnion bearings, that we had not previously included, and the total amount of steel required emerged as 1,200 tons, an increase of nearly 250 tons.

Toward the end of April we included all of these details and the increase in cost of many other items in an application to the DSIR for another grant. The £335,000 of our 1952 application had now risen to £439,000. The Nuffield Foundation would agree to provide only another £32,000, and it was clear that the bulk of this increase would fall on the DSIR budget.

The chances of getting this extra money looked poor, and as the DSIR advised us to seek other means of support, I began to wish that I had never wanted to build this radio telescope. Everything was in suspense. Blackett, too, became very annoyed and blamed me for introducing so much delay that the cream of the research would be skimmed elsewhere in the world with less ambitious apparatus.

Rainford, who had made independent contact with the finance officer of the DSIR, thought it was eager to help. In my previous encounters many of the members of the committee had been my wartime colleagues and friends trying to help and full of enthusiasm for the project. But committees change, and to my consternation I found an array of new faces around the table when I faced the DSIR committee on 6 May. A. M. Wilson, whom I had never met, had replaced Dee as the chairman. Sir George Thomson, now a member of the committee, with the support of others, no doubt anxious to make their mark with the DSIR hierarchy, immediately seized on the wind-loading problem. Within minutes we had retreated to the arguments of years ago—about wind tunnel tests, effects of icing, and all the various stability and aerodynamic problems I thought we had left far behind.

I was instantly attacked by G. P. Thomson about the safety of the structure. Backed by the other committee members, he questioned whether the correct allowances had been made for wind pressure on the mesh and other parts of the structure. They had, in fact, seized on the essential point. They did not for one moment question the structural calculations of Husband, but they asked for an assurance that the wind pressure assumptions on which the structural calculations had been based were correct. The assurance I could give them—the data from the wind tunnel tests on a sample mesh made at the National Physical Laboratory—did not impress them one bit. That was a flat surface, and I was asking for a 500-ton curved surface suspended 180 feet above the ground to withstand the forces of nature.

That was a quite horrible afternoon. Of the multitude of committee meetings I had attended, this one stood unique—and was to remain distinguished from all the meetings that lay ahead. The committee eventually decided that it wanted an independent assessment of the wind-loading issue from an aerodynamics expert, and the one piece of good fortune from that afternoon was the suggestion that W. A. Mair should be asked for an assessment.

I have already mentioned Mair as a university colleague whose advice I had sought about wind problems in the early days of the evolution of the idea of the telescope. Subsequently he had moved to a professorial appointment in the University of Cambridge, and now I was to reap the benefit of our earlier friendly association, when, against a phalanx of deans and professors who did not understand modern science and engineering, we had eagerly supported one another. When I repeated all this to Husband, he was greatly displeased, but we soon agreed that he would write a short report on the wind pressure assumptions for Mair's consideration.

There followed the quickest resolution of a major issue that I have ever encountered. Four days later we met Mair on a Sunday afternoon in the garden of his Cambridge house. It was one of those enchanting afternoons that only an English May can produce. The double cherry in Mair's garden was more redolent of "the double-blossom wild cherry-tree" in George Meredith's *The Egoist* than of wind pressure calculations. There Clara Middleton had turned her face to where the load of virginal blossom was "whiter than summer cloud on the sky." Now the wind was a gentle zephyr that dappled the shade on Husband's report, and it was hard to imagine the steady wind of 134 miles per hour that would be necessary to destroy the telescope—or of the worst conditions of ice and snow for which Husband had made these calculations. Mair signed the report, and I was soon on my way to London, waiting for the doors of the DSIR to open on Monday morning.

Although G. P. Thomson had been the leader of those who queried the basis of the calculations, he graciously withdrew his opposition when I wrote to him with a copy of this report. He replied that he was entirely satisfied, that he was sorry to have given me the extra trouble, but that for my own sake it was "as well to have the report signed by an aerodynamics expert just in case anything happened." With that release from anxiety I awaited the outcome of the DSIR Advisory Council meeting on 20 May with renewed confidence. However, my joy at its favorable response evaporated when I learned of its concern at the large increase in cost and of its warning "In the unfortunate event of more [money] being required it may be necessary for the University to raise it by appealing to another source."

I was not altogether surprised at Rainford's reaction, but dismayed by his instant response that the job could not possibly go on under those conditions. I felt trapped in a spider's web. Although we now had authority to spend the extra money, Rainford refused to sign the contract for the steelwork, because of the DSIR's conditions. Meanwhile, Husband had extracted a promise from United Steel that, if we placed the contract within a few weeks, it would complete the erection of the telescope by September 1954. We had reached early June 1953 and the coronation day of Queen Elizabeth II. Early that morning we heard that Edmund Hillary and Tenzing Norkay had reached the summit of Mount Everest. At least for me that helped place the telescope problem and the manipulations of recent weeks in perspective.

Another month passed before the arguments about the details of the contract with United Steel were settled. The price of steel continued to rise, but at last on 18 July all parties were in agreement. Throughout, the DSIR officials had shown much sympathy to me in these increasing frustrations. Although they were not entirely satisfied with some details of the contract, the chief finance officer told Rainford that they would not hold up the placing of the contract, because no doubt "Lovell must be very worried." Confidence returned, and as we looked at the last loads of concrete being poured into the ring beam Husband said that at last everything was in order and that the telescope would be erected the next year. It was a confidence that was not to outlast that summer of 1953 for long.

16

The Building of the Telescope: The Later Years of Construction, 1954–1957

By mid-March 1954 the double railway track had been fixed onto the concrete ring beam and leveled, ready to await the huge superstructure that was to rotate on it. The details of the central pivot and the azimuth bogies had been completed and contracts placed in 1953. No steel erection could begin until the pivot and bogies were in place, and our concern was that the production of these items would delay the steel erection. The pivot arrived on-site in May 1954; although one of the bogies arrived in July, the deliveries of the twelve were not completed until the spring of 1955, but by that time we were submerged in so many other problems that the delay in delivery was of little consequence.

One of these major troubles concerned the driving motors. In chapters 12 and 13 I described our early contacts with the Admiralty Gunnery Establishment, which led us to Metropolitan Vickers at Trafford Park. All seemed to be straightforward. On two of the bogies under each tower, 50-horsepower electric motors would be used in a servo system to rotate the telescope in azimuth, and two 50-horsepower motors on each tower geared to the gun turret racks and pinions would drive the bowl in elevation. The rotary converters needed to transform the alternating current of the supply into the direct current required by these motors would be placed inside the diametral girder immediately above the central pivot. All of this was eventually accomplished, but not in the straightforward and simple arrangement with Metropolitan Vickers that we had anticipated.

The severe problems about this driving system began to emerge in the autumn of 1953. On the advice of Metropolitan Vickers we had entered a sum of £28,600 in the estimates submitted to the DSIR on which the grant was based. In December 1953 we were dismayed to learn that this price was between £50,000 and £60,000—that is, double the sum we had allocated in the grant.

At that time a director of the company was Willis Jackson, who had been a professor of electrical engineering in the university. I suggested to Rainford that we visit him to explore whether the price could be reduced to the original estimate or whether the company might make a gift of this driving system to the university. I went to Trafford Park with Rainford on 21 December. It was a bad prelude to Christmas. Even on a summer's day the smoke and grit pouring from the chimneys of Trafford Park would obscure the sun. On that day of the winter solstice the place bore a resemblance to my idea of hell. I could hardly imagine that a mechanism for exploring the starry sky would ever emerge from that place—and it never did. Willis Jackson argued that we must have changed our specification if the company's price had doubled in two years, but it soon became clear that the original price was an irresponsible guess on the part of Metropolitan Vickers. Jackson not only failed to help us but even compounded our difficulties by making an approach to Lockspeiser at the DSIR, without our knowledge, with the suggestion that the DSIR should increase the grant. Lockspeiser refused to entertain the idea and told Jackson that "Manchester would never get another penny out of DSIR."

It was now the end of January 1954, and I knew that, in addition to the doubling of the price of the driving system, the prices quoted for the mesh and several other items were in excess of the estimates, but almost immediately we were faced with other troubles of such magnitude that these anxieties became a minor part of a far greater problem. The erection of the cranes did not begin until 23 March 1954, and I could have abandoned the project. We had then spent or committed about £100,000. I could think of no means of further savings and could have faced the DSIR with the unpleasant fact that, unless the grant could be increased once more, the telescope would not be built. I did not for one moment contemplate any such action, because by that time there were too many people around who would have been quite happy to abandon the project, and that would certainly have been the official response. Instead, I sought another source of money, and I must now explain the circumstances.

The Air Navigation and Air Warfare Committees

In the euphoria of returning to the university after the war, it never occurred to me that circumstances would once more associate me with the problems of armed conflict and defense. I submerged myself in academic affairs and became an astronomer. I retained some contacts with the research establishment and various ministries, but these were almost wholly concerned with securing equipment for use at Jodrell.

After a few years these relationships began to change. The Aeronautical Research Council, then reporting to the minister of supply, had a number of committees. One of these was the Air Navigation Committee, and in the early 1950s I was pressed to join this body. My initial reluctance to do so was overcome by the Astronomer Royal (Sir Harold Spencer Jones), the chairman of this committee. He had been so helpful and well disposed toward me that I could not reject his plea that there were problems on which my help was needed. My interest in military affairs was revived, and soon I was appointed to the Air Warfare Committee and then to the Aeronautical Research Council.

Thus, in the years of the construction of the telescope of which I now write, I became involved once more in the problems of warfare. These interests and contacts soon made me realize that the telescope I was building to study the heavens could have an influence on other, more sinister affairs. In the early months of 1954 I formed a scheme that made me confident that the Ministry of Supply would contribute substantially to the cost of the telescope and so overcome all the anxieties about the rising cost.

On 23 February 1954 a discussion of my memorandum on the potential of solar and cosmic emissions for navigation led to a recommendation that the ministry support research at Jodrell on centimetric emissions from space. That was all I had expected from that meeting, but a far more important event occurred as I walked out of the building into the London street. An occasional wartime contact now working in the Air Ministry had been at this meeting. Would I come with him to the Air Ministry to see Cockburn? Bob Cockburn had been in charge of the countermeasures division during the war in TRE, and we had lived through those years in adjacent laboratories. After a few years at the Atomic Research Establishment he had gone to the Air Ministry as its scientific adviser. He had recently been appointed principal director of guided-weapons research in the Ministry of Supply, but on that evening he was still at the Air Ministry.

Cockburn was already immersed in the formidable problem of defense against guided missiles, and he wanted an assessment of the performance of the telescope on various wavelengths and the cost of the modifications to give it a high efficiency on a wavelength of 10 centimeters. I discussed the prospects with Husband, and a week later Cockburn had my memorandum asking for support to make the necessary changes, on the grounds that it was in the national interest to modify the telescope for work in the centimeter wave band. The main modifications proposed were the reduction in size of the reflecting mesh to 0.75 inches, a stiffening of the bowl framework, the use of 34,000 bolts with site drilling and fixing instead of the less accurate bolting arrangements already specified, and an increase in the rigidity of the center tower holding the primary feed at the focus. The extra money needed to do this was estimated to be £46,000. At that stage I did not prejudice the ministry's interest by referring to the problem with the finance of the driving system and made a straightforward request for £46,000. In exchange I promised that the measurements the ministry required would be carried out and that if the ministry needed the telescope for "other purposes," this could be the subject of a separate arrangement.

The initial reaction from the controller of research in the ministry was favorable, but, for reasons that have remained obscure, when Cockburn arrived in his new post difficulties began to arise. In mid-April, Cockburn closed the matter in a formal letter, stating without explanation that with regard to the possibility of improving the surface of the telescope he could not justify giving a large sum from the defense budget. I wrote to Husband on 26 April to explain that I could not authorize him to engage in any modifications to the steelwork that would increase the cost over the present estimate. It was too late. He had already issued the drawings to United Steel with the strengthened back girder, and by this time I was so convinced of the national importance of the telescope that I did nothing further to stop this alteration. Subsequently it was easy and convenient for the committees of inquiry to accuse me of a failure to reveal these events. In fact, I was inhibited from giving an explanation; in any event, at that time neither Husband nor his chief structural engineer could estimate within 10 percent the actual tonnage of steel in this complex structure. An error of 10 percent was within the budgetary estimates for the steelwork, and so we proceeded to design and build the strengthened telescope.

The Ballistic Weapons

At the end of July 1954 the discussions in a meeting of the Air Warfare Committee gave rise to anxieties of a different kind, which were soon to interact with the problem of the telescope. There was evidence that both in the Soviet Union and in the United States effective progress was being made in the development of ballistic weapons. But at that point the United Kingdom lacked even radars capable of detecting supersonic bombers at maximum range. In early August I journeyed to my wartime research establishment, TRE, and discovered a largely negative attitude to these long-range-detection problems.

In this way another vitally important phase opened in the telescope story, and by mid-August a proposal for a joint program between ourselves and TRE was formulated with the request that the Ministry of Supply provide the balance of £30,000 needed to place the contract for the driving system of the telescope. Whereas in the March–April negotiations with Cockburn I had not raised the question of the driving system, the roles were now reversed. It seemed that within the 10 percent contingency for the steelwork we would have a stiffened and more accurate bowl, and the outstanding problem was our ability to drive it.

In the historical context the sequence of events during the next few months seems absurd. The authorities in the ministry, who really had the responsibility to initiate action on the radar detection of ballistic missiles, were passive in their attitude, and the inertia of those summer months began to undermine my will to continue. Lockspeiser was in a different category. He was a distinguished scientist, and his interpretation of the public duty of the head of the DSIR was that he support new scientific developments—and he did his utmost to do so. When I acquainted him with this situation, he gave me two pieces of advice. He could do nothing immediately, because of the promise extracted from him by the Public Accounts Committee that the DSIR give no more money for the telescope. "Wait until the spring," he told me, "when you really know how much more money you want to finish the job, and in the meantime tell the defense people to come and see me."

This was the autumnal equinox, and it seemed a long time to wait for the spring to which Lockspeiser referred, but at least I could take instant action with the defense people. Before the end of the month Cockburn had received yet another memorandum from me, this one dealing specifically with the telescope's possible role in detecting ballistic weapons. I asked him to visit Lockspeiser, and with an alacrity I had scarcely envisaged he did so with Sir Harry Garner, chairman of the Guided Weapons Advisory Board, on the last day of September. An up-to-date assessment of the extra money needed to complete the telescope was requested. By mid-October the DSIR staff had all the information they needed, and once more optimism returned when Cockburn told me he was certain we would be given the extra money.

Lord Simon and Sir Hugh Beaver

I believe that the Ministry of Supply would readily have made a grant to cover the deficit in the autumn of 1954. Unfortunately the ultimate source of money for the ministry and the DSIR was the Treasury, one of whose cardinal principles was that no project could be financed through two different avenues. The money would have to come through the machinery of the DSIR. The administrative officials of the DSIR responded that they had committed all their money for the next five years and had no more available for the telescope. The wear and tear of working through the official machinery began again, and quite soon the September stimulus of the Lockspeiser-Garner-Cockburn axis disappeared in the mechanism of grant giving.

I was constantly in the offices of the DSIR, and at last, on 17 December, I called once more with apprehension. The advisory council had met, and I had little hope that its members would wave the magic wand over the DSIR finances. I was wrong. The council had decided that every effort should be made to complete the telescope and had instructed the DSIR to find ways and means of providing the extra £40,000 to £50,000 that we then deemed necessary. Once more I had a stroke of good fortune in that Sir Hugh Beaver had recently become the chairman of the advisory council.

Sir Hugh came into my life through the influence of Lord Simon, of whom I wrote in chapter 14. Lord Simon, then chairman of the university council, and his wife, Sheena, were formidable influences in the political and educational life of those years. Lord Simon was never happier than when handling major projects and, although leaving me in no doubt about his opinion of the administrative muddle I had created, was determined that the telescope should be completed. At this stage in 1954 one manifestation of this determination occurred with great regularity on Sunday afternoons. The procedure formed a set piece. The Simons nearly always had weekend guests. By Sunday morning they would have determined whether their visitors would be interested in the work at Jodrell or might be able to help. There would be a telephone call; my wife would hastily pack a basket with provisions for tea; and we would drive to Jodrell and await the Simons, who would swoop like eagles with their weekend prey.

In this way Sir Hugh Beaver was drawn into the tentacles of Jodrell. The first occasion was on the Sunday afternoon of 3 October. Sir Hugh was quite unconcerned about rising costs. Like Lord Simon, he was well acquainted with major projects. During the war he had been director general of the Ministry of Works, and he calmed my fears by remarking that no engineering job of the magnitude of the telescope had been built since the war without the price's rising by several times. The Beaver, as he became known in the family circle, was to be a frequent visitor, but no other occasion had the significance of that first visit in October 1954 when he had just assumed the chairmanship of the advisory council.

The Growth of the Deficit

I do not doubt that, if the £40,000 to £50,000 had been the measure of our requirements to finish the telescope, our troubles would have ended with that meeting of the DSIR Advisory Council in December 1954. Unfortunately, when the problem might have been settled in this way as the year 1954 ended, we were no longer able to give an assurance that an additional grant of £40,000 to £50,000 would be sufficient for us to complete the telescope. The problems were centered on two difficulties—the driving system for the telescope and the design of the bowl of the telescope. The eventual solution of these two problems led to the official inquiries that came within a hair's breadth of ending my career.

The Driving System

I described earlier in this chapter our meetings with Metropolitan Vickers when we hoped to place the contract with that company for the driving system toward the end of 1953. Meanwhile Husband, who was naturally as irritated as we were with the attitude of Metropolitan Vickers, had been urgently investigating other possible means of control and supply. On 26 September 1954, when I was immersed in the contacts with Lockspeiser and with Garner and Cockburn from the Ministry of Supply, Husband told me that he had visited Siemens in Germany and that this firm would be able to supply an alternative driving system more cheaply and with far better delivery dates than Metropolitan Vickers.

Siemens was so anxious to obtain the contract that it offered a series of reductions in its price and delivery dates. By contrast Metropolitan Vickers not only increased its estimate further but also announced that it could not possibly deliver until September 1957. Although we made further efforts to negotiate with Metropolitan Vickers, hoping that it would react to competition from a foreign firm, we were soon forced to abandon the attempt. J. G. Davies had gone to Siemens to make sure that the servo system it proposed would be suitable, but it was early March 1955 before we could be certain that by placing all the orders for the system, including the gearing assembly, with Siemens we would save £18,000 over the Metropolitan Vickers estimate of nearly £60,000 and, moreover, could obtain delivery in nine months instead of the three years promised by the British company.

At last I felt that we were emerging from this almost inextricable tangle, but within weeks any hope of placing orders with Siemens or any other firm seemed to have vanished. First, Lord Simon and Sir Raymond Streat, the university treasurer, reacted fiercely when they learned we were about to abandon Metropolitan Vickers for a foreign firm. A few years later, when the British government was trying to ingratiate itself with the European Community, the opportunity of placing this order with a German firm would have been seized swiftly. That time had not yet arrived. In addition the hope that we would save £18,000 in this way was dashed by the news that there would be an import duty of 20 percent on this foreign machinery. That alone would have erased the projected saving; even worse was Husband's statement that proper provision for various ancillary items such as cabling had not been included in the original estimates.

The original figure in the estimates of £28,600 now bore the ugly and uncertain look of £65,000 to £76,000. But that contributed only partly to the development of the crises that made it impossible for us to take any further action on the driving system for more than a year. At last, on 16 May 1956, we finalized the documents accepting the Siemens tenders and sought the formal authority from the DSIR to proceed. By 23 June the orders had been placed not with Siemens but with the British firm of Brush. I have never discovered what pressures and intrigues during those few weeks of May and June in 1956 led to this last-minute switch from Siemens to Brush. Some person in high authority was so opposed to the pro-European movement in Britain that pressure was placed on Brush to quote a competitive price against Siemens. On 20 June, when the documents were about to be dispatched to Siemens, a telephone call from Brush quoted a price that would be less than the Siemens figure if import duty had to be paid and a satisfactory delivery date.

Within days Brush had the order. I was just thankful that this devilish issue had at last been resolved. In November of that year we met the chief executive of Brush and the directors of the firm of Wiseman, which was to make the gearing. They promised to deliver all the machinery and gearing by the spring of 1957, and they did so. By the end of May this driving system, which had been like a millstone around our necks for many years, appeared on the site as if by magic. As I write now, with the sun setting for the last time on a year thirty years later, I can hear the hum of those motors precisely controlling the motion of the telescope as it probes the heavens.

The Redesign of the Bowl

The crisis that led to a year's delay in placing the order for the driving system was caused by the design of the bowl of the telescope. I have already described the two quite separate issues that led to the protracted discussions between Husband and myself and the action to improve the shape of the paraboloidal bowl and reduce the size of the reflecting mesh. When I became concerned about the potential of the telescope in the ballistic weapons field, the anxieties were primarily about securing the extra money needed to purchase the driving system. Throughout the summer months of 1954, though, I became increasingly bothered by the rising costs of many items related to the steelwork—to such an extent that in May I decided that an access tower, moving on the diametral girders so that we could gain access to the focus, was not essential. We abandoned this and thereby saved some £5,000 to £10,000. At that time the design of the bowl structure was as it always had been from the beginning. The elevation drive was to be carried to the bowl through the gun racks and pinions. These huge gun racks, on each of the towers, were to be built onto a massive back girder. This would form a cradle for the relatively lightweight steel framework of the reflecting bowl. This back girder cradle would bear all the major torsional and other stresses. Unfortunately, the deepening of the back girder to give it more stiffness had led to unacceptable increases in weight. In July, Husband himself became concerned. The increased weight of this back girder made it essential to strengthen other parts of the design, and by the end of July it had become evident that this type of back girder design would make the part of the telescope that moved in elevation unreasonably heavy.

In mid-August, Husband told me that he had been forced to the conclusion that this back girder design was unacceptable. In later years Husband published a detailed account of these involved design features that led him to abandon the back girder in favor of a

space frame form of construction with a very deep circumferential girder. Sixteen main radial trusses were used to connect the panels of the circumferential girder to a central hub. The calculation of stresses in the framework was tedious but reasonably determinate, bearing in mind that the whole structure was supported at two points only. The most difficult parts of the framework to design were the members connecting the trunnions to the circular structure. The shear forces in the frame-work due to dead weight and wind forces acting at all directions through 360° were troublesome to deal with, and required a much greater weight of steel than had been allowed for in the preliminary structural estimates.*

It is easy to conclude retrospectively that at this point in mid-August I should have sounded the alarm bells to all concerned with the financing of the telescope. I might well have done so if at that moment we had known that this new design would require a much greater weight of steel, as Husband stated in his account years later. On 11 September, when I discussed the revision with Husband's chief structural engineer, I was assured that the weight was less than that of the back girder design, and three days later, at Jodrell, Husband said they had now computed the weight of the new bowl framework as being 520 tons. This was very little more than the original weight and significantly less than the 700 tons of the stiffened back girder.

We were eventually to be faced with a huge overexpenditure on the telescope. The subsequent private inquiries tended to concentrate on an obvious point that was not necessarily the vital one. When the files and records were searched, the investigators found an apparently alarming situation in which I had agreed privately to a "change in design" with Husband. It is true that I had at that moment agreed to a change that, to the onlooker, altered the appearance of the bowl of the telescope, but to me as a scientist it was a modification enabling Husband to meet our design requirements, which had been accepted in the 1953 revision. Years of experience have shown that, in fact, it did just that and no more. I agree that I was guilty of a technical administrative error of judgment in August 1954. However, it was not this error that led to the ultimate overexpenditure. This arose partly from issues that still had to be revealed and partly from the difficulties of estimating the cost of the large part of the structure that was in no way affected by this business.

*H. C. Husband, *Proceedings of the Institution of Civil Engineers* 9 (1958):65–86.

1955—A Year of Despair

Less than a year later I was to be accused by a committee of inquiry of agreeing privately with Husband to change the design not once but several times. The change from the back girder to the space frame design of the bowl mentioned above was one. Another, the consequence of which began to develop at that April meeting, concerned the reflecting surface of the paraboloidal bowl. We had already decreased the size of the mesh in a part of the surface to facilitate the use of the instrument on the 21-centimeter wavelength of the neutral hydrogen emission. During the autumn months of 1954 I had agreed to another change—the replacement of the open mesh by a solid steel sheet.

The initial stimulus arose from the difficulty of making good mechanical and electrical connections across the thousands of joints of the mesh. The manufacturer of the mesh could produce the mesh size and gauge that we required without difficulty, but the product was made in strips, as long as desired, though only a few feet in width. It would have been easy to fix these strips to the framework of the bowl, but no economical technique for making the thousands of joints across the strips when they were in position in the bowl could be devised. In the light of the experience then available we were asking for the joints to be as strong as the wire of the mesh and for an electrical conductivity across these joints as good as that of the conductivity of the material itself.

In September 1954 Husband first raised the question of using a solid skin made up of steel sheets welded together as a means of avoiding this problem. He would find the additional stiffness compared with the mesh helpful in the overall structure of the bowl. We were interested because there would be no loss of signal by leakage through the mesh. There was, of course, the problem of the wind forces. In the early design considerations the National Physical Laboratory had made extensive wind tunnel tests on various mesh sizes; as far as the ultimate safety of the telescope was concerned, Husband had made the assumption in calculating the wind forces that under certain conditions of snow and icing the mesh would effectively present a solid surface. The con-

cept of the solid sheet did not, therefore, present any new anxieties about the ultimate safety of the structure. On the other hand
we were worried that at operational wind speeds the solid membrane would be subjected to forces that, although not structurally
dangerous, might lead to distortion of the bowl away from the
paraboloidal shape and thus hinder our research work.

On 4 November 1954, when a decision about this was becoming urgent, I was lecturing at the National Physical Laboratory.
Afterward I saw C. Scruton, a member of their aerodynamics
division and an authority on the interaction of airflows with engineering structures. I told him of our concern, and he handed me
a report of his wind tunnel tests on a solid-sheet reflector he had
recently made for a commercial organization. This was much
shallower than the paraboloid we were building, and after some
discussion with Husband we decided to accept Scruton's offer of
making further measurements in the wind tunnel on a scale
model of the telescope. It is as well we did so.

The magnitude, and the largest out-of-balance forces—with the
elevation of the bowl at 60 degrees—were as already anticipated.
Unfortunately, that reassurance did not mark the end of the wind
tunnel tests. When I had first talked with Scruton at the National
Physical Laboratory in November, he asked me about the problem of oscillations of the type that had caused the destruction of
the Tacoma Bridge, in Washington State. Husband had always
been a little concerned that certain types of gusty wind conditions
might strip the teeth on the racks and pinions of the elevation
drive and had proposed to avoid this by allowing a certain
amount of slip in the pinions and racks in a high wind. Scruton's
final tests on the scale model in early January had been planned
to find out how the structure would behave in such gusty conditions.

I learned the grave results of these tests on a Sunday afternoon
early in 1955. In later life, as in my youth, Sundays became days
of relative peace. There would be church, the walks among the
trees in the garden in the summer, and the warmth of the fireside
in the winter. In my middle years Sundays did not hold this
release. In the war Sundays were always days of work and meetings, and then came the years of the telescope about which I write

now. In the end some of these Sundays were to hold exciting and memorable events, but in the 1950s the memories are of despair and frustration. One such Sunday was 23 January 1955.

The bad news from Husband on that Sunday afternoon was that he had just returned from the National Physical Laboratory, where Scruton had completed his wind tunnel tests on the model of the telescope. In this model springs had been used to simulate the racks and pinions by which the bowl structure was to be driven from the tops of the towers. With gust simulation violent oscillations had occurred, to such an extent that in a mean wind speed of only 40 miles per hour the springs had been shattered. Husband's background worry about the stripping of the teeth on the racks and pinions had suddenly emerged as a major problem.

A year later it was convenient for the inquisitors to point to my agreement to change to the solid membrane as the primary cause of the expensive modification needed to eliminate this danger. It was idle for me to argue that this was not true, that the same destructive oscillations would have occurred with an iced-up mesh. No doubt today the designers of a Tacoma Bridge or of a telescope have computer programs that cover all such interactive effects between nature and the structure, but in those days we were working in unknown territory. The item first materialized at that meeting on 5 April in Husband's cost estimates under the heading "braking system." Ultimately Husband proposed to solve the problem with a "bicycle wheel" fixed to the perimeter of the bowl framework and bearing on large tractor wheels, mounted on the diametral girder. This slender wheel bore no load, but the friction on the tractor wheels damped out any incipient oscillations of the bowl.

At the end of that meeting on 5 April, which began so optimistically in the belief that we could complete the telescope for about £70,000 extra, of which the DSIR had promised £50,000, we did not know how much the overexpenditure would be. The final steel tonnage was uncertain, we did not have a price for the manufacture and fixing of the solid membrane, and the seemingly endless series of minor items was continually escalating the price. We agreed to suspend all further orders for any parts of the system until a full assessment could be made of the probable cost.

Paramount was a more accurate estimate of the steel tonnage now required, and when we parted from Husband he agreed to produce these data in a few weeks.

It was five months, not a few weeks, before all the necessary data had been assembled that would make it possible to assess the real overexpenditure. One serious cause of the delay was the difficulty in finding a contractor who would undertake to manufacture and weld the steel plates on the framework of the paraboloid. We spent the summer months measuring the shape and testing the conductivity across plates made by a firm that eventually gave an estimate of nearly £50,000. This was nearly three times greater than we had estimated in the 1953 revision of the costs and was so unrealistic that Husband eventually found a firm that promised to carry out the work for a sum nearer to the estimated cost. Although the structure continued to grow, it was to be a long time before the fixing of the reflecting surface of the paraboloid became possible.

During the autumn of 1955 our true plight began to be revealed. In September and October, with closer estimates of the tonnage of steel, it was clear that a serious underestimate had been made. I had been prepared to learn that the weight might have exceeded the estimates by a hundred tons or so, because of the introduction of the bicycle wheel damping system, but was entirely unprepared for the mid-October figure of 1,660 tons—an increase of 440 tons. Even worse was the claim now being made by United Steel for a price of £135 per ton, as against the £114 on which the estimates had been based. The catastrophic result was that the complete telescope was now estimated to cost £630,000. Even with all the money that had ever been promised by the DSIR we were nearly £250,000 in debt.

The officers of the university were in an anxious state, but when the chief finance officer of the DSIR came to Jodrell to assess the position, near the end of October, he gazed at the thousand tons of steel rising above the Cheshire plain, turned to me, who was anxiously awaiting the expected reprimand, and remarked, "The strength of your position, Lovell, is that huge mass of steel." Maybe, but the DSIR had only £50,000 held in reserve for us, and on that afternoon the finance officer held out the hope that the DSIR Advisory Council might be persuaded to make a pound-for-pound offer on the remaining £200,000 if the university would raise £100,000.

1957—The Year of Drama

The cumbersome and tortuous nature of the government decision-making process for once worked marvelously in our favor. The storm that was around us did not break over our heads until March 1957. By that time I knew that, whatever would be my fate, nothing could stop the completion of the telescope. After the visit of the chief finance officer of the DSIR in the autumn of 1955 the great steel structure of the telescope grew majestically over the Cheshire plain. It was like a fantasy. With a debt that was now more than a quarter of a million pounds, and increasing daily with the rising cost of steel and labor, the work continued. Although there was no money to pay for this work, all the contracts had nonetheless been placed and nothing was actually held up for the lack of money. By mid-September 1956 the entire framework was complete. The bowl was ringed with steel, supported by sixteen towers containing ninety miles of scaffold tubing. The huge cranes had finished their work. In the autumn sunshine the filigree of steel was a stupendous and unforgettable sight.

Although the bowl was still encased and supported by the towers of scaffolding tubes, Husband was anxious to measure the azimuth track friction. Toward the end of January 1957 the workmen began to cut away the supports on the tops of these scaffolding towers, so that the trunnion bearings at the top of the two towers of the telescope took the weight of the bowl. The scaffolding tower was released from the central hub of the bowl on 1 February, and two days later on a quiet Sunday afternoon the final supporting members were removed. There was not yet any means of driving the telescope mechanically, so a number of 50-ton hydraulic jacks were applied to the bogies. Only an inch of movement on the azimuth tracks was necessary to make the measurement, but the telescope had moved and a dream began to be a reality.

At that moment the telescope was like a skeleton, without a full reflecting surface or any means of driving. The contract for the supply and the fixing of the steel sheets on the framework of the bowl to form the reflecting membrane had been placed with Orthostyle. Although the main framework of the bowl had been completed by mid-September 1956, much minor riveting and fixing of the supports delayed a start on the membrane until mid-November. By 23 December only 80 feet of the 250-foot-diameter surface had been completed. The Orthostyle workmen then abandoned the task for Christmas and refused to return until the winter months had gone. They did not begin work again until April, and their strikes, union activities, and "working to rule" gave me an unpleasant glimpse of the troubles at the core of the British labor force. At last in May the plated area of the bowl began to increase once more.

On 12 June 1957 the motors first rotated the telescope under power. I was not there to see it. On the first day of that month, with the telescope proceeding inevitably to completion, I had gone with my family to rest in North Wales a hundred miles away. Although I had returned temporarily for a meeting a few days later, I did not know precisely when Husband would decide that all was ready to take this great step. The twelfth of June was a perfect day for the mountains, and I had climbed Snowdon with the three elder children. At the end of the day we returned to the sea shimmering in the evening sun. Excited faces awaited us—Husband had telephoned from Jodrell to say that the telescope had moved in azimuth under power. The next morning the postman delivered a letter from my secretary that reported, "We have just seen the telescope move 20 yards—it was most exciting and the suspense was terrific! In fact it is now being driven back to its original position. Everything seemed to go very smoothly."

Throughout that summer there had been a few marvelous occasions when for brief periods the troubles were forgotten. I had been absent on 12 June when the telescope was first driven under power in azimuth. Nothing could have torn me away from Jodrell eight days later. When I was a young man my summer days were dominated by the game of cricket. On the morning of an important match I would be awake as the sun rose, anxiously looking at the sky to make sure the weather would be good. Day-break on 20 June 1957 found me transported back to those early memories. It was to be a vital day that once again I desperately wanted to be calm and dry—and now cricket was far from my thoughts. This was the day assigned to the attempt to drive the bowl of the telescope away from the zenith for the first time. On this critical matter hung the designer's fate and all of my hopes for the future. The days of midsummer in England can be wet and windy, or they can be days of surpassing beauty. On this June day of 1957 I saw the globe of the sun rising through a thin mist and, hearing the joyful song of the skylark high above my head, knew that this crucial test was only hours away.

Husband must have been traversing the Pennines at that early hour, for when I arrived at Jodrell he was already supervising the clearance of the last pieces of the scaffolding towers. At ten-thirty all was in readiness. My wife had arrived with a friend, and we were in the shade of the ancient oak just outside my office door. There was the hum of the motors, and suddenly a new sound— the screeching and distress calls of hundreds of starlings that had discovered that many parts of the structure were convenient for nesting. They had sensed the motion before we could see the huge bowl slowly tilting away from the zenith. As I walked onto the structure, I had the sense of walking on the deck of a great ocean liner moving through a calm sea. The hum of the motors, the gentle vibration, all in good order and under perfect control. On that day the troubles that engulfed us were forgotten. The only worry was whether the birds would return to their young in the nests.

When the bowl of the telescope was first driven away from the zenith, on 20 June, the plating of the reflecting membrane was still not completed. The steel plates were about three feet square and were fitted and welded ring by ring, proceeding outward from the center of the bowl. On 20 June the twelve outer rings remained to be fixed. By the end of July the whole paraboloid was surfaced, and we prepared to make our own crucial test to discover how this huge structure, which had taken five years to build, would perform as a radio telescope. There was not yet any means of remote control, nor were the position indicators working, so for these first recordings we marked the azimuth track with degrees in chalk and similarly the stablizing bicycle wheel for elevation. The telescope had to be driven "by hand" from the noisy motor room on the diametral girder, and there was much shouting and running around throughout those August nights. The first recordings on 2 August were of the radio emission from the Galaxy, with the bowl remaining in the zenith. The recording was of no particular scientific importance, but for us it was an inked trace on a paper chart symbolizing an end to five years of massive engineering and a beginning to the researches that we had planned and dreamed of for many years. During the next few nights we measured the polar diagram of the reflector by driving the telescope from the motor room so that the beam swept across the radio source in Cassiopeia. We were delighted with the results—the beam shape and power gain of the paraboloid were exactly as we expected, and everything was fine.

The Intervention of the Public Accounts Committee

The excitement and pleasure as the telescope moved to completion in those August days of 1957 were overshadowed by a deep crisis that had begun to arise after the visit of the chief finance officer of the DSIR to Jodrell Bank in October 1955. Then he had made the suggestion of a pound-for-pound contribution from the DSIR if the university would raise a half share of the deficit. The officers of the university were naturally in a state of alarm, and I had been summoned before them and instructed to raise this money, then believed to be about £100,000. Nevertheless, throughout all these troubles and those that were to follow, these senior officers never for one moment lost their faith in the project. There were storms enough within their offices, but always the demand was that I should finish the telescope and raise the money to cover the deficit. This was a remarkable corporate act of faith that was to bind me to them and to the university for the rest of my career.

Unfortunately, the straightforward but burdensome pound-for-pound offer did not materialize in the manner we had expected.

On 5 December we learned that the DSIR Advisory Council had demanded that there should be an inquiry with an independent chairman. This committee of inquiry met on 18–20 January 1956. As the days of that winter lengthened into spring, we heard that the committee had recommended to the advisory council that the telescope be completed on the basis of the pound-for-pound arrangement. By that time the estimate of the final deficit had increased to £260,000, and the measure of our greatest happiness would have been to receive the official advice from the DSIR that it would find one-half of this sum, leaving us to raise the remainder. Indeed, we did eventually receive such advice. It came in mid-August, but because the Public Accounts Committee (PAC) had already taken an earlier interest in the finance of the telescope, Treasury approval had to be obtained. The Treasury agreed to this further contribution but with a series of conditions, occupying four pages of typescript, to which neither the DSIR nor the university could possibly agree.

The worst of the conditions was that the university give a written guarantee, supported by the opinion of counsel, that the money now needed was not greater than it would have been if legal action for the recovery of the money had been taken. This was a grievous insult to the university and to all who had been associated with the telescope. Even today I am amazed that at that moment in August of 1956 the building of the telescope was not abandoned. Fortunately, Husband was unaware of this condition, and for their part the university officers reacted by instructing me to explore alternative avenues for raising the money.

Although the unacceptable conditions imposed by the Treasury had been conveyed to us in August 1956, it was not until March 1957 that the storm broke. A year earlier there had been an unconfirmed rumor in a Sunday newspaper about the debt we had incurred, but this did not lead to further publicity. The situation changed with dramatic suddenness early in March 1957 when the report on the DSIR inquiry of January 1956 was published by the comptroller and auditor general at the Treasury, Sir Frank Tribe. The report was factual and concerned the increasing cost of the telescope, without any explanation of the circumstances. The British press of 6 March seized on these facts, expressing righteous indignation at this needless waste of the tax-payers' money. The public commentators had been critical of me, but their main target was Husband, who was naturally furious at the publication of the financial details without any explanation of the reasons.

No public rejoinder was possible. The facts in the Tribe report were correct, and in any case the whole matter was sub judice because of the tense legal aspects involving the university and the DSIR with the Treasury. In fact, the Tribe report was merely the official prelude to the investigation by the PAC on 21 March. It was unusual for the PAC to concern itself with such minor financial issues as the debt on the telescope. It was unfortunate that during this period there was increasing agitation in political circles about the freedom from public accountability enjoyed by the universities in Britain, although they were financed extensively by public money. For those who were striving to gain public accountability the finances of the telescope provided a fine opportunity to illustrate that universities were incapable of controlling the expenditure of large amounts of public money. Even so, since the next investigation of the PAC dealt with the unauthorized expenditure of more than £30 million by the army and the air force, the telescope problem of a mere £260,000 might well have seemed unimportant and the issue would have ended with a reprimand had there not been a serious issue concerning the evidence given to the PAC.

Lockspeiser had unfortunately retired under the age limit, and a vital thread of contact had disappeared. He had been succeeded as secretary of the DSIR by Sir Harry Melville, a distinguished professor of chemistry in the University of Birmingham. It was Melville who had the unenviable task of appearing before the PAC on 21 March to answer questions about the telescope. In later years I was to become closely associated with him on the grant-giving bodies, but at that time in March 1957 we were strangers. A month before the meeting of the PAC, Melville had made a brief visit to see the telescope, but his limited familiarization with the history of the instrument led to a number of grievous misstatements of fact when he was interrogated by the members of the PAC.

On the afternoon of 13 August the report of the PAC, with a verbatim account of the interrogation of Melville at its meeting on 21 March, was published. Although we had been fearful of a critical report, none of us anticipated the extent of the personal attack on Husband and myself that now appeared in print. Husband was accused of having altered the design without consulting me and I of irresponsibility in using public funds.

For an engineer to be accused of lack of cooperation was naturally a very serious matter for his firm and his livelihood. Since we had been together almost every day for five years, it is hard to see how Melville could have given such false information together with erroneous statements about my living in apartments on the site. Nevertheless, there it was in print in an official government publication, and Husband's fury was boundless. On the evening of that day he demanded that I write a letter to *The Times* of London denying these allegations; otherwise he would sue me for damages. I consulted Rainford, who said that I could not possibly write such a letter to *The Times,* because the PAC report was a privileged document. In that case, said Husband, he would drag the information out of me in the witness box.

In that manner, at this moment of accomplishment symbolizing years of close partnership between engineer and scientist, I was advised to speak to Husband only in the presence of the university solicitor. Melville and the PAC had reduced a great engineering and scientific endeavor to the level of a charade. I could not, and still cannot, understand how such eminent civil servants and members of Parliament could allow such a report to be published in a protected document without checking that their information was correct or taking the trouble to see for themselves what was involved.

The thrill of the first probings of the universe and the realization of the dream of years were thus enveloped in a near disaster that, but for a miraculous occurrence, would have ended my career and landed me in prison.

17

The Impact of Sputnik

On 24 August 1957 I left these troubles behind me to attend the meetings of the International Scientific Radio Union (URSI) at Boulder, in Colorado. One of the commissions of URSI was specifically organized to discuss the results and techniques emerging from the new radio investigations, and the presentation of the various results from Jodrell was a main reason for my attendance. Two occurrences there seemed to me to mark a turning point in world affairs. The first was a news item in the *New York Times* of 27 August giving the text of an announcement made by Khrushchev:

> A super-long-distance intercontinental multi-stage ballistic missile was launched a few days ago. The tests of the rocket were successful. They fully confirmed the correctness of the calculations and the selected design. The missile flew at a very high, unprecedented altitude. Covering a huge distance in a brief time, the missile landed in the target area. The results obtained show that it is possible to direct missiles into any part of the world.

Two years earlier, on 29 July 1955, President Eisenhower had announced that the United States would launch an earth satellite during the International Geophysical Year (IGY). A similar announcement was made a day later by the Soviet Union. The world believed that the United States would do this but dismissed the announcement of the Soviets on the grounds that they could not possibly have the technical capability. Shortly before I left England a scientific colleague had returned from Moscow and informed me that the Soviets had an earth satellite nearly ready for launching. When I saw the announcement by Khrushchev in the *New York Times*, I realized that the Soviet promise to launch an earth satellite for the IGY was about to be fulfilled, using the launching rocket of their intercontinental ballistic missile.

The second occurrence followed almost immediately. In a packed lecture theater I listened to John Hagen, the director of the U.S. Project Vanguard, lecturing on this American proposal to launch an earth satellite. He said that the development had been delayed and that it would be some months before Vanguard could be launched. After the meeting I told Hagen of my belief that, if this were the case, the Soviets would be the first to launch an earth satellite. He dismissed the idea, apparently sharing the general view in the United States that the Soviets had no such technical capability.

I was greatly disturbed. President Eisenhower had decreed that the American project should not be based on military technology. Vanguard was based on sounding-rocket technology and was outside the mainstream development of the U.S. military ballistic rockets. As Hagen was to reveal in the aftermath of the tragic Vanguard failures, there were inadequate financial arrangements and a degree of priority barely appropriate for the IGY programs. The Soviets had no such inhibitions about separating their IGY program from a military basis.

These developments seemed to me to be indicative of a rapidly changing balance of power in the international scene, but it did not occur to me that they had relevance to the troubles I faced at Jodrell. It is true that we had a provisional arrangement to use the telescope to track the Vanguard satellite in connection with our measurements of the electron density in the earth–moon space, but a delay of months would be of no consequence to us.

When I left Jodrell on 24 August, I hoped and expected that by the time of my return the seemingly endless series of miscellaneous clearing-up jobs on the telescope would have been finalized. They were not. In mid-September I was dismayed to discover that the rate of completion was slowing down almost to zero. In particular, no progress had been made in the linkage of the telescope drive to the control room, and we seemed far away from having a completely steerable telescope.

My handwritten diary of those days is before me as I write now. The entry for 2 October 1957 epitomizes the scene:

> The telescope atmosphere has been exceedingly disturbed since my return. First, I was infuriated at the impotence of the Site Committee. . . . Husband has continued to find every possible excuse for not doing things. At the Site Committee [on 24 September] he said that 14 days intensive work was required to link up with the control room. After 7 days [the] Brush [engineer] appeared and succeeded in driving it backwards. They were due to come again on Tuesday but . . . there was a blow up because their Chief Engineer refused to come back with "University people breathing down his neck." This led to a vicious exchange between Husband and me about this and several other matters.

Eventually the Brush engineers were persuaded to return, and by the evening of 1 October the telescope was being driven satisfactorily in azimuth from the control room computer. That evening we were able to make some measurements with the telescope driven in right ascension automatically and scanned by hand in elevation. Our practical and financial troubles were "leavened by a few hours use of the telescope"; the automatic drive was phenomenally good, and that diary entry ended with the comment that the telescope was "a wonderful sight in the evening sky."

As I wrote that on 2 October, I could not have suspected that this was the penultimate entry in a diary I had been keeping regularly for more than five years. Three months elapsed before I had time to make a final note. Then, on New Year's Day of 1958, I wrote, "Two days after the last entry, that is on October 4, the Russians launched their first earth satellite. All life changed immediately." At the beginning of October I had remarked to a colleague that only a miracle could raise us out of the bottomless pit of our troubles. Within a few days the launching of *Sputnik I*, the world's first artificial earth satellite, provided us with a vision of the miracle that we could grasp.

The news of the Soviet success swept the world as no other scientific and technical achievement had ever done. *Sputnik I* was far heavier and larger than the proposed American Vanguard satellite. It epitomized, in a form for all to witness, the dramatic advance of Soviet science and technology in the years since the end of World War II. In particular the scorn and disbelief that had commonly been poured on Soviet pronouncements by America almost instantly turned into a total reevaluation of the U.S. effort.

The Telescope and Sputnik

At Jodrell we had developed a method whereby it was possible to measure the total electron density in the earth–moon space from a study of the radar echoes from the moon. It seemed to me that if we could use an earth satellite as an artificial moon close to earth, then by comparison with the lunar results we ought to be able to measure the electron density in the interplanetary space between the satellite and the moon. It was an innocent suggestion made with little awareness of the appalling difficulties of carrying it out—or of the quite unforeseen consequences. The important point in the history of those days is that there was an official proposal, as part of the Jodrell Bank IGY effort, to carry out this experiment using the telescope as a radar device to follow an earth satellite. I regarded it as a rather minor and speculative part of our IGY program.

On 4 October I had no intention whatever of using the telescope to observe *Sputnik I*. Contractors were in possession, we could not drive it from the control room, there were few research cables on it, and the radar apparatus, which we intended to use eventually on the American Vanguard, was not on the telescope; neither had we any immediate plans for installing it. We could, of course, have used the telescope to receive the beacon transmitter from *Sputnik I*—the "bleep-bleep"—but this seemed pointless because these signals could easily be picked up by a cheap commercial receiver connected to a small dipole aerial.

For reasons I did not understand, the news media in Great Britain seemed to think that the telescope was associated with

Sputnik—or ought to be. On the morning of 5 October (a Saturday) the avalanche of interest was merely beginning, but I was surprised to find a number of reporters at Jodrell. Throughout the Saturday and Sunday a state of siege of newspaper and broadcasting personnel began to develop around my home and Jodrell. I am still uncertain why this happened.

I suppose that honor and the press would have been satisfied if we had connected a receiver to the telescope and steered it to receive the bleep-bleep from the satellite, but since the bleep-bleep was coming anyhow from simple receivers on the ground at Jodrell, I refused to engage in this kind of pointless demonstration for effect. For me it was either the radar experiment already planned for the telescope or nothing, and with our apparatus and the telescope in the condition they were at the time any rational judgment would have placed the possibility of doing this experiment many months away.

On Monday morning the situation changed dramatically, for two major reasons: one concerning my own outlook and the other involving Husband. As far as I was concerned, a number of telephone calls from various places in London informed me that there was no defense or other radar in the country capable of detecting the carrier rocket that had placed *Sputnik I* in orbit. In chapter 16 I described my involvement in 1954 with the Ministry of Supply in the attempt to gain its financial assistance with the telescope on the grounds of its potential value in relation to ballistic weapons. Subsequently, as the telescope problems absorbed my time and energy, I had been out of touch and simply could not believe that those responsible had so neglected the situation. Before the end of that day I had learned with incredulity that, at least in the free world, not a single radar had succeeded in locating the carrier rocket—and this was the rocket of a Russian ICBM! The miracle was waiting to be grasped. With the vast gain of the telescope even the low-power radar transmitter we were using on the moon experiments should be able easily to obtain a radar echo from the carrier rocket.

The installation of this lunar equipment in a few days would have been a formidable task even if the telescope had been fully operational and under our control. Husband had informed a

259

member of the press over the weekend that the telescope was "operational," and he had expressed considerable annoyance to me that, as far as he and the contractors were concerned, we had not demonstrated its potential by receiving the satellite signals with it. When the situation changed on Monday, I told him that my attitude toward the bleep-bleep remained the same but that we would attempt the radar experiment if he, for his part, would summon all the parties necessary to complete the work so that the instrument could be driven through the computer from the control room.

Work that had previously been thought to take months was then completed in forty-eight hours. The Brush engineer who had refused to continue with "University people breathing down his neck" now returned and completed the job with university people and the whole pack of press people breathing down his neck. The change in attitude under this stimulus was fantastic. At 6 P.M. on Wednesday, 9 October, the servo loop to the telescope drive was closed, and for the first time the instrument was moving automatically under remote control from the control room.

Meanwhile we had to face enormous scientific problems consequent upon the decision to attempt the radar experiment. The entire equipment, weighing many tons, was in a laboratory hundreds of yards from the telescope. To move it bodily onto the telescope at short notice was out of the question. Our only hope was to connect it by transmission lines across the intervening field and then by flexible cable on the telescope itself.

Our first ambition was to test the telescope with this lunar transmitter connected to it, to see whether we could obtain radar echoes from the moon. By the Wednesday evening the telescope was moving automatically on the correct coordinates, and a few minutes after midnight the telescope was transmitting to the moon and strong echoes were received on the cathode-ray tube. The entire place was seething with reporters, live television was being transmitted, and the noise and glare of light and publicity made any scientific task ten times more difficult. At that stage the BBC alone had many more engineers on the site than I had on my whole staff at Jodrell Bank. They captured the excitement and the first radar echoes obtained with the telescope.

ASTRONOMER BY CHANCE

Echoes from the moon satisfied us that the equipment and the telescope were performing satisfactorily, but we had far to go before we could hope to achieve a radar contact from a fast-moving carrier rocket whose position in the sky was only vaguely known to us. On top of all of these problems I faced a severe split among my own staff. It happened to astronomers all over the world. Some were enthusiastic about the opening of this new age; others regarded it as an interruption to their work. The former immediately saw new possibilities for scientific research; the latter complained that the vast sums of money involved would be better spent on more telescopes and equipment. So it was in the local Jodrell community. I do not for one moment blame those who retired to their homes or offices disgruntled that we had temporarily taken the telescope away from the slow and painful processes of initiating its researches in the conventional way, but I am profoundly grateful to the half dozen who stood by me in those historic days. They remained imperturbable and highly efficient under this extraordinary pressure. It was in J. V. Evans's small laboratory, almost monopolized by the cathode-ray tube display, that we saw the first lunar echoes during the Wednesday night. With this lunar apparatus we planned to locate the rocket, but our success with it was short-lived. Almost immediately some obscure fault developed that—probably because of the extreme weariness of all concerned—could not be located.

At this moment on 8 October a crucial encounter with J. S. Greenhow took place. Greenhow had joined us as a research student in 1948 and had developed the meteor radar technique to measure the height of the ionized meteor trails and to study the ionospheric winds that distorted them. The transmitter he used most frequently worked on a wavelength of 8 meters and, with a peak power of 150 kilowatts in the pulse, was one of the most powerful of these small transmitters that we had in use at Jodrell. Once more I was walking on the path to Evans's laboratory, and Greenhow was coming toward me, a piece of paper in his hand—a torn envelope containing, I thought, a problem about his upper-atmospheric winds. No! The pencil marks on the back of the envelope were to convince me that if we could place his small meteor transmitter in the elevated laboratory, suspended underneath the bowl of the telescope, we would be able to locate the carrier rocket. His proposal had two important advantages. First, the transmitter would be fed through only about seventy feet of cable to the dipole at the focus of the bowl, and thus we could eliminate the heavy losses of power in the long cable and transmission line connection to the more powerful lunar transmitter. Second, his transmitter worked on a frequency of 36 megahertz, compared with the 120 megahertz of the lunar transmitter, and hence the transmitted beam would be nearly four times broader, giving us a far greater chance of success in the search for the carrier rocket.

With this transmitter and receiver, designed for research on meteors, we searched for the rocket during the night of 11 October. The cathode-ray tube was full of the radar echoes from meteor trails, and, although there was absolutely nothing to guide us as to what an echo from a rocket would look like, we were reasonably satisfied that one of the responses was at such a range and of such a character as to be from the rocket. The next evening (Saturday, 12 October) there was no doubt at all. Just before midnight there was suddenly an unforgettable sight on the cathode-ray tube, as a large fluctuating echo, moving in range, revealed to us what no man had yet seen—the radar track of the launching rocket of an earth satellite, entering our telescope beam as it swept across England a hundred miles high over the Lake District, moving out over the North Sea at a speed of 5 miles per second. We were transfixed with excitement. A reporter who claimed to have had a view of the inside of the laboratory where we were wrote that I had leaped into the air with joy. I cannot remember whether this is correct, but I do remember turning to my elder son, Bryan, who had begged to be present in a corner of the laboratory and who was at that moment showing more interest in the arts than in science at school, with the words ''If the sight of that doesn't turn you into a scientist, nothing ever will.'' He did eventually become a geologist.

By this time it seemed to me that the eyes and ears of the whole world must be on Jodrell Bank. We had no press officer, no public relations officer, or any means of dealing with this situation. However, within these few days the entire atmosphere changed, and the friendliness of all of those present enabled us to survive and succeed in this complex public and technical situation. It was nevertheless impossible for me to answer questions endlessly day and night, and, at the suggestion of the Press Association, I agreed to give two press conferences every day, at 6 P.M. and 11 P.M. Night after night, both before and after our first echoes, our lecture room was packed with reporters and cameramen. After two weeks of this life the pressure slackened, but only temporarily. The launching of *Sputnik II* on 2 November with the dog Laika on board stimulated the interest to even greater heights. By this time the lunar equipment had been repaired, and with this the rocket of *Sputnik II* was detected when it was over the Arctic Circle at a distance of a thousand miles. With this equipment we also observed the final orbits of the carrier rocket as it burned up in the atmosphere in the early hours of 1 December.

On 29 October in the House of Commons, Prime Minister Harold Macmillan said, "Hon. Members will have seen that within the last few days our great radio telescope at Jodrell Bank has successfully tracked the Sputnik's carrier rocket." By any rational human standards our troubles should have been at an end.

The Aftermath of Sputnik

If our troubles had been merely financial, a solution would, I think, have been close at hand in that fantastic autumn. Jodrell Bank and the telescope were scarcely out of the news for months. We began to gather public sympathy and enthusiasm as the telescope became a symbol of at least some British contribution to the "space age." Placards appeared, issued by the government, featuring the telescope as "Britain's great achievement." Letters descended on MPs by the hundreds, and there was lobbying at the highest levels.

Unfortunately, issues of deep principle had become involved, as well as money. False statements damaging to Husband were in the public record; in turn he had threatened me and the university with legal action for damages. The Treasury had presented the university with unfulfillable conditions even for the DSIR to make an additional grant for the telescope, and the promises the Public Accounts Committee extracted from the DSIR and the Treasury made it impossible for the government to settle the bill and close the matter honorably, however much it may have been politically desirable to do so.

The bleak contrast between, on the one hand, the public acclaim and, on the other, the inability of the responsible bodies to make progress toward freeing the telescope for our use exasperated me so much that I tendered my resignation to the vice-chancellor. He summoned me to his office, where he gently but firmly restrained me from any such impetuous action. I returned to Jodrell; on my desk were two messages, one from the Academy of Sciences in Moscow thanking me for all the information we had sent them about the tracking of the rockets, and one from Sir Charles Renold (the chairman of the site committee) informing me that I would not be allowed to use the telescope until the legal position about the handing over of the telescope to the university had been clarified.

Before any progress whatsoever could be made, two major issues had to be settled. First, some formula had to be found by which the Treasury's conditions respecting the award of the additional grant to the university by the DSIR could be met. Second, the wrong done to Husband and his firm by the erroneous statements in the PAC report had to be redressed. The two were interconnected and tangled. The university could have met the Treasury's conditions by moving legally against Husband. It was clear that a single step in this direction would have been instantly countered by Husband's suing me for damages for allowing the PAC's published record to stand.

The nadir was reached on a Sunday afternoon, 9 February, a day of leaden skies and bleak wind when not even the emerging snowdrops could lift my spirit. Sir Raymond Streat, who had succeeded Lord Simon as chairman of the university council, telephoned to ask if he could see me. He arrived at The Quinta, our house in Swettenham, at teatime as the grayness of the day was merging into darkness. As he spoke, I began to despair. His personal sympathy and kindness could not, and never did, remove the sting of the message. The university council had given him the unpleasant task of conveying to me its opinion that Husband was about to serve me with a writ for damages of a million pounds. They believed that as long as the PAC record was allowed to stand, the university had little hope of defending the case and that without the means of paying the million pounds the consequence would be my imprisonment.

This final catastrophe did not happen, because the inexhaustible patience, skill, and common sense of those involved slowly but surely found areas of agreement through which progress could be made. The vital step was the withdrawal of the previous evidence before the PAC. To this end an important move was made by Melville at the end of February when he wrote to the clerk of the PAC conveying a memorandum from Husband in which the factual evidence of our long and close association was summarized. Whether or not this committee would accept Melville's assurances and publish a rebuttal of its original accusation—the only condition under which Husband would abandon his legal action—remained unknown. The vice-chancellor and Rainford were summoned to appear before the committee in March, and neither they nor the officers of the university were hopeful of a successful outcome.

As the days lengthened, we slowly moved away from the lip of the precipice. On 24 July 1958 the PAC presented its report: "In view of this further evidence, it is clear that the evidence given to the Committee of last session was gravely inaccurate and misleading, and that there was in fact the fullest collaboration on scientific and technical matters between the consultants and the University Professor."

The greatest single impediment to action was removed. The university was at last released from its shackles and moved into action. With the receipt of the additional £130,000 grant from the DSIR the immediate target was to clear the remaining capital deficit of a similar amount. There was the additional anxiety of a claim by the United Steel Structural Company for £124,405 in excess of the figure on which the overall deficit was based. Fortunately, this great additional burden was avoided by the generosity of the company.

By the autumn of 1958 the entire atmosphere had changed. The telescope was again involved in international affairs of great moment, and I was at last free to seek financial aid from the DSIR. On 28 October 1958 Lord Simon came to Jodrell. He came frequently, and this may not have been his last visit, but although I saw him at his home on several occasions subsequently, I like to remember that day as his farewell to the telescope (he died two years later). As usual, his black pocketbook came out, and with pencil poised he asked the question frequently posed in the past. What were my troubles? There were none except the money. I was not to worry; that would sort itself out. "But technically what about the telescope?"—"Oh, that's marvelous." He snapped the pocketbook closed without a note and said, "Good, I *do* like success."

18

Lunik and *Pioneer 5*
—The End of Debt

The events described in chapter 17 at last freed the telescope for our use. Once more I was able to obtain research grants from the DSIR, but there remained the problem of raising the £130,000 from other sources to clear the remaining debt on the telescope. At the end of November 1958 the university launched a public appeal for £150,000 "to free the instrument from debt and thus enable all available resources to be used in furthering the research work for which the telescope is designed."

Our expectation that the continued success of the telescope and the publicity surrounding its engagement in the Soviet and the American space programs would make it easy to raise this money turned out to be quite wrong. The large money-raising parties held at Jodrell during the spring and summer of 1959 were rather embarrassing events. Many of the major industrialists took the view that the telescope had brought credit to the country and that therefore the government should clear the debt. Lord (Quentin) Hailsham, then minister for science, was annoyed by the avalanche of letters descending on him to this effect. These well-wishers missed the point entirely, which was that after the clearance of the legal problem the government, through the DSIR, had honored its agreement to pay half the outstanding deficit and that the university was now attempting to raise the balance independently to fulfill its share of this agreement.

The unfortunate result of this somewhat obtuse attitude on the part of those from whom we expected substantial contributions was that by July 1959 the fund had reached only £65,000 and then increased by only a few hundred pounds by the end of the year. Fewer than a hundred industrialists had responded, with relatively token sums; the bulk of this money had been sent by private groups and individuals. Schoolchildren sent us their pocket money, and my files are full of touching letters from mothers of young children, disabled people, and pensioners making real sacrifices in order to subscribe a few shillings to the fund.

The emotional response was overwhelming. We did not lack public interest and goodwill. The newspapers did their best on our behalf, repeatedly drawing attention to the irony of the debt. But we still lacked the money we urgently needed and by the end of 1959 were still £85,000 short of the sum required. In the public eye the telescope had become a symbol of British scientific and engineering achievement, but it was burdened with debt— depressing to the spirit of those of us who lived with it, a millstone on our scientific researches. However, we were soon to become involved in a series of remarkable American and Russian space launchings that brought an end to our debt and anxieties.

The First American Moon Rockets

In the spring of 1958 I received a clandestine visit from an American air force colonel who informed me that the U.S. Air Force (USAF) had decided that it could use its Atlas ICBM to launch a rocket to the moon; he added that the USAF could be ready by August but had no means of tracking it. Would we use the telescope for this purpose? This request, which was the reason for the colonel's transatlantic flight, suddenly appeared before me like the vision of another miracle that, if it came to pass, would surely mark the end of our troubled times, and we were soon talking of the detailed arrangements. A complete trailer load of tracking equipment would be flown over from the Space Technology Laboratories in Los Angeles, about one month before the launch, together with a small party of technicians.

For three months the arrangement made in my office remained a complete secret. Apart from a few members of my staff it would have remained so until the day of launch but for an eagle-eyed reporter who noticed with surprise one day in July a large trailer on the road approaching Jodrell Bank, with the container plainly marked "Jodrell Bank, U.S. Air Force, Project Able." The next day, 25 July, the *Manchester Guardian* ran the headline "Jodrell Bank Joins in Journey to Moon" and commented, "That the United States Air Force should have fallen back on the use of Jodrell Bank for this purpose is a telling tribute to the versatility and power of the great telescope."

These early attempts of the USAF to send a rocket to the moon failed in their main purpose but brought an avalanche of favorable publicity to Jodrell Bank and the telescope. In the first attempt, made on 17 August 1958, the Atlas rocket exploded eighty seconds after leaving the launch pad and did not rise over the Jodrell horizon. In the second attempt, on 11 October, the probe traveled nearly 80,000 miles into space and then fell back to earth. At the final burnout after launching the velocity of the probe was 34,000 feet per second instead of the required 35,250, and the probe eventually burned up as it reentered the earth's atmosphere over the South Pacific. Even so the Jodrell telescope had tracked the probe and recorded the data from the scientific experiments on board exactly as planned.

Although Pioneer failed as a moon rocket, it was a striking scientific success; in particular it produced the first information about the nature of the zones of trapped particles and the magnetic field out to a distance of nearly 80,000 miles from earth. The events of that weekend were eventually to lead to Jodrell Bank's salvation.

The Soviet Moon Rocket, September 1959

Less than a year after these events with the USAF Pioneer, and with the American contingent still in the trailers near the telescope, we were involved in the drama of the Soviet moon rocket.* When the announcement of the launching was made from Moscow, I was playing cricket.

After the cricket match I returned to Jodrell in the evening to find the Americans in a state of frenzy. They had been harassed the entire afternoon by officials in Washington asking and demanding that they attempt to locate the rocket. Unfortunately, they could reply only to the effect that I was playing cricket and that nothing could be done until I returned. I was unmoved by this interest from across the Atlantic and remained in a casual frame of mind until I discovered that there was a message from Moscow on the Telex machine. This presented precise details of the frequencies of the transmitter in the Lunik and the exact coordinates calculated for the latitude and longitude of Jodrell Bank, giving the time of lunar impact as 10 P.M. BST on the following day (13 September 1959). Since the telegram had been dispatched from Moscow less than one hour after launching, it was obvious that the Russians had prepared the calculations for Jodrell Bank in advance and that they had an almost audacious confidence in the success of the launching.

We sprang into action. Fortunately, the aerial on the telescope being used in the research program covered the Lunik frequency band. To save time we carried a portable communications receiver to the swinging laboratory and, suspended underneath the bowl, we tuned in without difficulty to the bleep of Lunik and found it to be on the precise coordinates given in the Moscow telegram, and at that moment 100,000 miles from earth. The predicted impact time was 10:01 P.M. BST the next day, a Sunday. The excitement and the tension as the time approached remain vivid in the memory. At 10:01 the bleeps were still loud and clear, at 10:02 we began to think that it might have missed, but twenty-three seconds later the bleep ceased—the first man-made object had reached the moon!

The news monopolized the front pages of the national news-

---

*This was played out in less than thirty-six hours, for the first six of which I played cricket. I described my nonchalant reaction to the news of the launching and my awakening to the realities of the situation during the cricket match in an essay "The Moon Match," in Michael Meyer, ed., *Summer Days* (London: Eyre Methuen, 1981), pp. 123–31.

papers on the following morning, 14 September, and they carried their own stories of those hours at Jodrell Bank. At that stage in the history of space science and technology—only twenty-three months after the launching of the first Sputnik—the Lunik mission was, of course, an impressive achievement. The event came on the eve of Khrushchev's visit to the United States, and the political significance of the conjunction of these two events was widely noted. The Americans were once more trailing, and to my amazement there were signs of disbelief. Vice President Richard Nixon said, "None of us know that it is really on the moon." A former U.S. president, Harry Truman, said the feat of the Russians was "a wonderful thing—if they did it."

Fortunately, during the last hour of the rocket's flight J. G. Davies had measured the Doppler shift as the rocket fell under the gravitational field of the moon, and this measurement made it possible to calculate the general region of impact. He was able to do this by comparing the frequency of the received signal on the 19-megahertz equipment with a standard in the control room. It was an absolutely crucial measurement, which left no shadow of doubt that impact had occurred. But for this, many Americans would, I think, have continued to doubt the reality of the achievement. Some weeks later the director of the American moon program remarked to me that Davies's measurement was a flash of genius. It was the first time such measurements had been made; the technique was to be refined and used in all subsequent moon experiments to establish the motion of rockets under the gravitational field of the moon.

The Soviets were very soon to circle the moon. *Lunik 3*, launched on 4 October 1959, succeeded in transmitting back to earth the first photographic record of the hidden side of the moon. The telemetry was received at Jodrell Bank, and the recorded tapes and other data were dispatched to the Soviet Union through diplomatic channels. In November I was happy to receive a graceful letter of appreciation from A. N. Nesmeyanov, the president of the Soviet Academy of Sciences. However, away from these epoch-making events, we still faced the cold, hard world of the debt on the telescope, and as 1960 dawned this seemed to me to be an incubus with which we would have to coexist forever.

271

Pioneer 5

It was our good fortune to be closely associated with the first striking American success in deep space, in the spring of 1960. *Pioneer 5* had originally been conceived as a probe to Venus, but after the failures of the lunar Pioneers *Pioneer 5* was to become a probe into deep space without a specific planetary objective. The spacecraft was built with a number of major scientific experiments in view, and there was every prospect of gathering rich scientific dividends if data from these experiments could be obtained as the probe traveled millions of miles into space.

Whereas in all our previous work with space vehicles we had acted in a "passive" capacity—that is, we had received the signals transmitted by the probe—on this occasion we were to transmit to the probe as well, in order to command its various functions. The Jodrell telescope was still the only instrument that offered any hope of transmitting with enough strength to the probe, over distances of tens of millions of miles, to command its functions.

In the history of Jodrell Bank, 11 March 1960 was an epic day. In addition to the command of *Pioneer 5*'s transmitters in space the telescope had the vital job of transmitting the signal to the probe to release it from its carrier rocket after it had been launched from Cape Canaveral. When Pioneer was 5,000 miles from earth, a touch on a button in the trailer at Jodrell transmitted to the probe a signal that fused the explosive bolts holding the payload to the carrier rocket. Immediately the nature of the received signals changed, and we knew that *Pioneer 5* was free, on course, and transmitting as planned. For the rest of the day the spacecraft responded to the commands of the telescope; when it sank below our horizon on that evening, it was already 70,000 miles from earth. The next evening it was beyond the moon. The signal transmitted by the telescope captured the imagination of the assembled newspapermen as day after day the telescope commanded and recorded the vital information from *Pioneer 5*, as it sped into the depths of the Solar System.

Throughout all the days and nights of our association with the Soviet and American space launchings, I was constantly sum-

moned to the telephone. We had no public relations or press officer, and I discovered that polite and attentive responses were
essential to the preservation of the excellent relations now established with the media. But the constant repetition of the same
information was a wearisome affair, and my voice must have
conveyed this feeling as once more I lifted the receiver a few days
after the launching of *Pioneer 5*. To my astonishment this was not
the voice of the media but Lord Nuffield's: "How much money
is still owing on that telescope of yours? I want to pay it off." I
tried to thank him, but he replied, "That's all right, my boy, you
haven't done too badly."

So we became the Nuffield Radio Astronomy Laboratories, in
recognition of the contribution made by the Nuffield Foundation
and by Lord Nuffield personally. In that way a fairy tale became
a reality, ending the years of anxiety the depths of which were
known only to my own family.

19

Astronomical Research: Radar Astronomy

The telescope was built for the purpose of fundamental researches into the nature of the universe. The publicity attached to Jodrell through the association of the instrument with the Soviet and American space launchings almost completely obscured this fact in the public mind. For another decade after the *Pioneer 5* episode in 1960 the telescope continued to play a significant role in these space programs. By the time Neil Armstrong and Buzz Aldrin landed on the moon, in 1969, both the United States and the Soviet Union had acquired their own comprehensive tracking networks, and our participation had decreased markedly. Although these international events absorbed a good deal of my own time and energy, few of the Jodrell personnel apart from J. G. Davies and a new technical recruit, Bob Pritchard, were involved to any great extent. Indeed, statistically the diversion of the Mark I telescope to this space work throughout the 1960s occupied a tiny percentage of its operational time; contrary to the belief of the media and the public, the operational time of the instrument was primarily absorbed in the astronomical researches. It is these aspects that I shall describe here and in chapter 20.

Research in fundamental science tends to be highly competitive, and in this new subject that we were developing, the interest

274

and activity began to spread rapidly in the 1950s. It would have been useless to have a large working steerable telescope in 1957 if Jodrell had not been able to remain in the vanguard of research. The techniques of observation were in a phase of rapid change, and observations of the radio emission from the heavens, and their interpretation, were in the process of changing our ideas about the nature of the universe. By the autumn of 1952, when the pile drivers began work on the foundations of the telescope, I had already recognized the danger of a major shift of interest of the research staff to the nonscientific problems of the telescope that were beginning to surround us. I could not escape such a major diversion, but it soon became an important duty to throw a protective screen around the research staff. Throughout, Hanbury Brown and J. G. Davies were the two from whom I constantly sought advice, but even in these cases the thrust of the discussions was nearly always on how to use the telescope and the plan of the researches as soon as it became operational. As year succeeded year with the telescope still incomplete, these plans continually evolved and changed. That they were able to do so, because of the important development of the Jodrell researches during the years of construction, is of paramount importance to the eventual success of the telescope.

During the years of construction of the telescope there was a substantial shift of emphasis in the research at Jodrell Bank. The interest in radio astronomy as distinct from radar astronomy was stimulated by the success of the transit telescope in detecting the radio emission from the Andromeda nebula and by important developments elsewhere in the world. Nevertheless, throughout the 1950s and in the early years of operation of the 250-foot telescope, significant research in radar astronomy continued at Jodrell, and examples of work carried out with the instruments operating in the radar mode will be sketched in this chapter.

The Meteor and Moon Radars

In the account of the work on meteors given so far, I have described the measurements of primary astronomical interest—

that is, the constant surveillance of the incidence of meteors on the atmosphere and the measurement of the velocities and orbits. At the same time we were concerned with the physical processes taking place when the small meteoric particle evaporated in the high atmosphere. Many scientific papers were published about these aspects of the research in the late 1940s and throughout the 1950s. One such topic concerned the behavior of the ionized trail of the meteor after it had been formed in the 100-kilometer region of the atmosphere. For trails whose radio echoes could be observed for more than about half a second, it was evident that peculiar distortions were taking place. It was quickly recognized that the fluctuations of the radar echo must arise because the long, thin column of ionization of the meteor was being distorted by ionospheric winds.

J. S. Greenhow was engaged on this research at the time of the launching of Sputnik. Under favorable conditions he could determine the wind vector at a given height from only thirty minutes' observation. By the mid-1950s he had collected a mass of data about the diurnal and seasonal variation of these winds in the 80- to 100-kilometer region. They could reach 100 meters per second and within an hour could reverse their direction. Furthermore, at points separated by only 10 kilometers, the wind speeds could differ by 50 meters per second. The publication of these data and their interpretation, in terms of thermal and tidal effects of the sun, occupied a great many scientific papers published by Greenhow and one or two other students during the years of construction of the telescope. Greenhow did not use large aerial systems; for most of his work he used simple dipoles. To these he connected one of the many ex-army radar transmitters that we had accumulated, and he designed and built his own specialized receiving and recording equipment.

Greenhow left Jodrell Bank in December 1959 to work at the Royal Radar Establishment,* by which time he had published more than two dozen papers. However, Greenhow is enshrined in my memory as the person who approached me on a day in October 1957, when I seemed to be besieged by all the demons

---

*Unfortunately, Greenhow died in 1962 at the age of thirty-four and at the peak of his career.

in existence, holding a torn envelope on which he had penciled a few figures. It was his equipment transferred to the telescope that changed our future in one dramatic night in 1957.

Although Greenhow's meteor radar equipment mounted on the telescope first detected the carrier rocket of Sputnik, it had been our intention to use the more powerful lunar transmitter for this purpose, as was described in chapter 17. Indeed, it was with the transmitter and receiver of this lunar radar equipment that the automatic following of the telescope was first tested shortly after the launching of *Sputnik I,* in October 1957. The original plan to transfer this lunar echo equipment in its entirety to the top of one of the towers of the telescope was implemented. With this equipment rebuilt to avoid interference with aircraft navigational systems, a systematic series of measurements of the electron density in the earth–moon space was made for several years. In June 1961 the work was extended in collaboration with a group at the Sagamore Hill Observatory, in Massachusetts. The transmissions through the telescope at Jodrell were received after reflection from the moon at both Jodrell and Sagamore Hill during the periods of mutual lunar visibility, and much useful information was obtained about the time scale of the irregularities in the ionosphere.

Planetary Radar

The ease with which lunar echoes could be obtained by means of the Mark I telescope soon stimulated us to far greater fields of endeavor, and we turned again to the calculations on the planets that I had done for Blackett in 1946. It was evident that a radar contact with the planet Venus presented an exciting prospect. Two major uncertainties lay behind all the accumulated optical data on that planet. First, the question of the distance of the sun from the earth was involved. This distance, commonly defined in the literature as the solar parallax, or the angle subtended by the radius of the earth at the sun, is of fundamental importance in astronomy. It is the baseline from which the distances of the nearer stars are measured and hence the distance scale for the universe. Many expensive expeditions had been made in an effort to improve on the observation first made by Edmund Halley in 1716 to determine the times of the transit of Venus across the solar disk at points on the earth's surface widely spaced in latitude. The net result of all these attempts to measure the solar parallax differed by one part in a thousand, even though the best of the individual measurements claimed an accuracy of one part in ten thousand. In other words, the fundamental unit of distance in the universe was uncertain by a tenth of a percent. Radar contact with the planet Venus would settle this matter without ambiguity, since the time between the transmission and reception of the radar echo could be measured with precision. This would give the exact distance to the planet, and so the earth–sun distance could be computed precisely by means of Kepler's laws—the orbital period of the planets around the sun being known.

The second issue concerned the rotation rate of Venus on its axis. The planet is perpetually cloud covered, and this had frustrated all efforts to measure how fast the planet was rotating—or even in what direction it was rotating. The current estimates, based on theoretical arguments, covered a whole range of possibilities, from a few hours to nearly a month. In this case the observation of radar pulses reflected from the planet would enable the differential velocity of approach and recession at opposite limbs of the planet to be measured.

Although the sensitivity achieved with the large telescope made the observation of radar echoes reflected from the moon an easy task, elementary calculations indicated the formidable difficulty of establishing a radar contact with the planet Venus. Even at close approach to the planet Earth, Venus is nearly a hundred times more distant than the moon. The received signal from a radar pulse scattered from the moon or a planet varies inversely as the fourth power of the distance. Thus the distance factor alone would make radar contact with Venus a hundred million times more difficult than contact with the moon.* Since the telescope could follow the planet continuously, the stratagem of integration

*The problem is eased to the extent that the radius of Venus is 3.5 times greater than that of the moon. The sensitivity required for radar contact with Venus is actually about 5 million times that required for radar contact with the moon. The factor is about 100 million for Mars and 3 billion for Jupiter.

could be used. That is, the successive radar pulses scattered from the planet could be superimposed in the receiver, in which case the signal-to-noise ratio would be proportional to the square root of the number of pulses superimposed.

At the close approach of Venus to Earth in September 1959 this technique was used with the telescope equipped with a powerful klystron transmitter (previously used by our colleagues in the university in a linear accelerator to produce high-energy protons). With the telescope tracking the planet, radar pulses were transmitted for a period of five to six minutes (the time of travel of the pulse to and from the planet). The transmitter was then switched off and the telescope used as a receiver for a similar period. Fifty-nine hours were integrated in this manner, and there seemed to be a barely significant increase in signal strength. The statistics were weak, and we would no doubt have kept the data to ourselves but for a curious and, as it transpired, misleading coincidence. In the United States the development of very powerful transmitters for defense purposes had led workers at the Lincoln Laboratories of MIT to attempt the radar contact with Venus in 1958 by means of a much smaller telescope. They obtained data similar to the Jodrell 1959 results, and although both observations were suspect, the results were published—showing a significant difference between the radar result and the best of the available optical measurements. By the time of the next close approach of the planet, in April 1961, we had improved the sensitivity of our equipment and no longer had any doubt about the radar contact. In America and the Soviet Union large teams with access to powerful transmitters, although with smaller telescopes, also achieved success.

All the results were consistent. In an average value of 149,600,000 kilometers the spread in the radar results from the various groups was only 2,500 kilometers. The 1959 results were shown to be spurious, but even so a large discrepancy of 60,000 kilometers between the radar value and the best value of the conventional optical values existed. The radar value of 149,600,000 kilometers was eventually accepted internationally as the correct value on which the solar parallax is based, and the discrepancy between it and the optical determinations, claiming such small errors of measurement, remains one of the minor physical or psychological puzzles of astronomical research.

The 1959 Jodrell result, like the 1958 American value, was false. Although the details of the 1959 Jodrell work were published with some caution expressed, the data obtained in 1961 nevertheless showed that their publication had been unwise and misleading. My deep suspicion of results claimed on the basis of low statistical certainty remained with me for the whole of my subsequent career. I was not a direct participant in the observations or their analysis, but no scientific papers were published from Jodrell without my scrutiny and consent, and that brief communication to *Nature* was a blemish in our record. It was a hard lesson to be learned that any desire for priority or publicity should never influence a rigorous scientific judgment, and to the best of my knowledge I do not think that a similar error occurred in any of the hundreds of research papers subsequently published on the work at Jodrell.

We could now concentrate our attention on measuring the rotation rate of this perpetually cloud-covered planet, and I have described in *Out of the Zenith** the circumstances that led us to do so in collaboration with Soviet scientists. The plan was to use the telescope and transmitter of the space probe tracking installation on the Black Sea coast to transmit to the planet and receive the signals scattered from the planet with the Jodrell telescope. In the event, there were endless communication and programming difficulties, the latter being as severe at Jodrell as in the Soviet Union. We at last succeeded with these bistatic observations on 9 January 1966, and the results revealed that the planet was in retrograde motion with a period of 243 days. But the information was no longer a surprise. At the close approach of June 1964 the Americans at the Jet Propulsion Laboratory, in Pasadena, California, and a group using the huge Arecibo reflector in Puerto Rico had preempted us in these measurements of the rotation of the planet. We could not compete on this massive scale using extremely powerful transmitters, the design of which had been stimulated by defense requirements. Neither I nor my colleagues had any desire to trail in these planetary radar researches, and it

*Oxford and New York: Oxford University Press and Harper & Row, 1973.

was not a difficult decision to abandon this line of research in favor of the researches in which our own telescope would be unique.

The Final Stages of Radar Astronomy at Jodrell

In 1959 and 1960 the telescope had been equipped with a powerful transmitter for defense purposes under the circumstances to be described in chapter 22, and some interesting radar observations of meteors and the aurora borealis were made and published. This work provided a useful cover for the real reason for the installation of the transmitter but did not have sufficient intrinsic interest for us to pursue it as a research project. In fact, the termination of the bistatic observations with the Soviet Union in the spring of 1966 marked the end of our radar studies with the Mark I telescope.

I witnessed the passing of the active radar researches at Jodrell with considerable regret and nostalgia. My hope and expectation that the new telescope would be used not just as a purely receiving radio telescope but also in the active radar mode survived for a few years only. However, it would have been foolish of me to oppose a transition made inevitable by various circumstances. The majority of the Jodrell staff and all the incoming research students were eager to use the telescope to study the radio emissions from the universe, for which it was an unique instrument. On the radar side we no longer had the money and manpower to compete with the radar systems developed in the USA and the USSR. Although these used smaller telescopes, they had access to extremely powerful transmitters built with defense interests in mind. Finally, a technical problem made it increasingly difficult to run a research establishment in which both sensitive receiving devices and radar transmitters were in use. Radar and radio transmissions are anathema to those using sensitive receivers. For a few years we had used a synchronized pulse cabled around Jodrell so that the receivers would be suppressed during the period of any transmitted radar pulses. As our receivers became more sensitive and the radar systems more powerful, this stratagem lost its efficacy. By the time the Mark I telescope was operational, the two techniques had become incompatible. If the telescope was used as a radio telescope, receiving the weak radio emission from space, it became impossible to use transmitters elsewhere at Jodrell, even on a widely separated wavelength. For a few years the telescope was thus used either as a radio telescope or as a radar transmitter and receiver for the lunar and planetary work, and the other radar researches at Jodrell (mainly the meteor work) closed down. Quite soon the transmitters and the radar equipment I had secured from my wartime colleagues as a gift, or for a token sum, were collected by the scrap merchants—the last relics of the years of mud and leaking cabins, which few now remember.

20

Astronomical Research: Radio Astronomy

The new telescope soon absorbed all parts of the radio astronomical researches in progress at Jodrell with smaller instruments. As well as the concentration of the work on the measurement of the angular size of the remote radio sources the telescope became a significant instrument for the detailed surveys of the radio emission from the sky and for many other activities. All these uses of the telescope were predictable, and since it first operated, in 1957, seven hundred scientific papers describing the results of this work have been published. Some of the details of the use of the telescope up to 1970 are given in my book *Out of the Zenith,* and since that time these researches have developed further, with the help of computerized systems and new types of receivers that could not have been foreseen in those first years of the use of the telescope.

I do not intend to write about these developments here. The survival of a modern research instrument for more than thirty years is unusual in this age. The adaptability of the telescope to the new techniques and its significance in the pursuit of unforeseen discoveries have been critical in its long life. As I write now, thirty years after the first, tentative measurements, the researches with the telescope cover objects and events in the universe that were unknown and unsuspected in those early days. A few of these, illustrative of the progress of scientific discovery, and exciting and all-absorbing to those involved, will be described in this chapter.

The Problem of the "Radio Stars"—Radio Galaxies

When we began to gain operational use of the 250-foot steerable telescope, in 1957–58, there had been a complete reversal of opinion about the nature and location of the radio sources. In chapter 10 I referred to the meeting of the Royal Astronomical Society in October 1951 where there was almost unanimous agreement that these localized radio sources were invisible radio stars in the local galaxy. An important discovery was made in 1952 when Walter Baade and Rudolph Minkowski used the 200-inch optical telescope on Mount Palomar to search for faint visible objects that might be correlated with the two strongest sources of radio emission. Their search was possible because in 1951 Graham Smith, working with Ryle's radio astronomy group in Cambridge, refined the radio measurements so that he was able to determine accurate positions for these two radio sources. When Baade and Minkowski used the 200-inch optical telescope to take long-exposure photographs of the regions of the sky in these positions, they discovered that the strongest of these radio sources, in Cassiopeia, coincided with a nebulous object in the Galaxy having the characteristics of a supernova remnant. On the other hand, the second-strongest radio source, in Cygnus, coincided with a peculiar extragalactic nebula estimated to be at a distance of about a thousand million light-years.

Within a few years the opinion strengthened that the majority of the radio sources were indeed extragalactic objects. Ryle and his team in Cambridge produced a catalog of 1,936 of these radio sources, and a study of the number-intensity distribution caused them to change their opinion that the sources were in the galaxy. Now they argued that these unidentified radio sources formed a class of rare extragalactic objects similar to the source in Cygnus. But the evidence was circumstantial since by the time the telescope became operational, in 1957, only seven of the radio sources then cataloged had been identified optically as abnormal extragalactic objects of the Cygnus type.

By 1960 these remaining doubts about the extragalactic nature of the unidentified sources had disappeared, primarily because of the series of measurements made at Jodrell Bank during the years when the telescope was being constructed. This research immediately occupied a substantial operational time of the new telescope, with profound consequences to our understanding of the nature of the universe. These consequences followed from the development of techniques that achieved a resolution in the radio spectrum comparable with that attained by the large optical telescopes. On a wavelength of a meter, the 250-foot-aperture paraboloid would have a resolving power of only a little under one degree—twenty times inferior to that of the unaided eye. A single paraboloid of this type working in the meter wave band would have to be 140 miles in diameter to approach the resolution of half a second of arc of the modern optical telescopes. This problem stimulated the radio astronomy groups in Sydney and Cambridge in the late 1940s to develop the idea of the radio interferometer.

As long as the telescopes are so close together that the signals can be fed along a good cable to a common receiver the problem is easy, but for the majority of radio sources the two aerials must be many miles apart in order to achieve sufficient resolution. Cable connections are then impossible, and the development of radio links, instead of cables, to transmit the signals from a distant radio telescope to the transit telescope at Jodrell Bank occupied a large part of our research effort from the early 1950s. We had the great advantage of the high gain of the transit telescope, and in the early years of this research a smaller mobile aerial was moved to points at successively greater distances from Jodrell. By 1956, with this mobile aerial 20 kilometers distant from the transit telescope, three of the radio sources in the zenithal strip of sky available to the transit telescope remained unresolved.

This could mean only that the three unresolved sources must be of the same type as the remote radio galaxy in Cygnus. Rudolph Minkowski had been our main contact with the optical astronomers, and he recognized the possible significance of identifying these unresolved radio sources. Using the 200-inch Palomar telescope, he concentrated on one of these—cataloged several years previously at Cambridge as 3C 295, lying in the constellation of Boötes. The conclusion that the angular diameter was less than 12 seconds of arc meant that this radio source was ten times smaller in angular extent than the Cygnus radio galaxy. It was also seventy times fainter at the radio frequency used in the Jodrell measurements. It thus seemed logical to conclude that the object was far more distant than the one in Cygnus.

By 1960 Minkowski had obtained photographs with the 200-inch telescope that showed the presence of a distant cluster of galaxies surrounding the position of the radio source. One of the brighter galaxies in this group was precisely in the position of the radio source 3C 295, and Minkowski had little hesitation in concluding that this was the galaxy responsible for the radio emission. He measured the redshift, and this implied a recessional velocity of more than 40 percent of the velocity of light and a distance of about 4,500 million light-years.

This clear demonstration of the belief that the small-diameter radio sources might be objects analogous to Cygnus, but at much greater distances, was instantly seen to be of immense significance in cosmology. The concept of radio galaxies was thus born: peculiar objects, at that time believed to be two galaxies in collision, at great distances in the universe. Compared with normal galaxies they were powerful emitters of radio waves and thereby endowed the radio telescope with the capability of penetrating far into space. At this moment of intense interest in the angular-diameter measurements that had led to the discovery of this remote radio galaxy in Boötes, we were able to use the 250-foot steerable telescope. Hitherto the use of the transit telescope had limited our investigations to the relatively small number of sources that could be detected in the zenithal strip. Now we had complete coverage of the northern sky, and during the next few years our major use of the telescope was on this work.

The Discovery of Quasars

In the early stages with the new telescope longer baselines were obtained by using the remote aerial at sites in North Wales, but a number of sources remained unresolved. The remote aerial was then moved to an airfield in Lincolnshire, and the British Broad-casting Corporation allowed us to mount a repeater system on one of its television masts high on the Pennines. Even during the North Wales phase of this work results of great consequence were obtained. The stimulus for the identification of the radio source 3C 295 by Minkowski had been the discovery that the angular diameter was less than 12 seconds of arc. Now seven out of the first ninety sources measured were found to have angular diam-eters less than 3 seconds of arc. The logical conclusion was that these must be radio galaxies of the same type as Cygnus and 3C 295 but at an even greater distance in the universe. The possibil-ity of obtaining optical identifications immediately assumed great importance.

The first news of this attempt at identification was given by Allan Sandage during the meetings of the American Astronomical Society held in New York City between 28 and 31 December 1960. In September he had used the 200-inch Palomar telescope to photograph the region of the sky in the position of one of these radio sources, cataloged as 3C 48. His announcement that a faint blue star existed in the precise location of 3C 48 and that it was accompanied by a faint wisp of nebulosity caused a sensation in the astronomical world. He concluded that this must be a rela-tively nearby star with most peculiar properties. Shortly after-ward Sandage identified two more of these small-diameter radio sources with similar starlike objects of a blue color. Astronomers were startled by these discoveries. The smallest-diameter radio sources yet measured, which all logical reasoning had indicated as being remote radio galaxies in the universe, were apparently nearby stellar objects in the Milky Way.

The great difficulty was that the spectrum of these objects could not be identified, and a veil of obscurity surrounded the subject until 1962 when observations by a former Jodrell Bank student, Cyril Hazard, using the radio telescope at Parkes, in New South Wales, gave the key to the remarkable nature of these objects. He employed the technique he had developed at Jodrell to study the occultation of one of the unresolved small-diameter radio sources known as 3C 273. The accurate position thereby determined led to an identification with another peculiar starlike object accompanied by a jet or wisp, on a plate taken with the 200-inch tele-scope. The issue of *Nature* for 16 March 1963 contained details of this work and of a comment by Maarten Schmidt on the nature of the "star." Schmidt had studied the spectrum and had con-cluded that the only explanation was that this was not a star in the Milky Way but a hitherto unknown type of distant extraga-lactic object. The same issue of *Nature* brought the even more remarkable news that, stimulated by Schmidt's conclusion that the "stars" were distant objects, Jesse Greenstein and T. A. Mat-thews had again studied the spectrum of 3C 48 and had con-cluded that the observations could be made to fit a redshifted spectrum of hydrogen lines if 3C 48 had an apparent recessional velocity of 110,200 kilometers per second, implying that it was one of the most distant extragalactic objects known. The enig-matic objects that had for over two years been believed to be stars in the Galaxy were, on the contrary, found to be among the most distant objects known in the universe.

The way was now open to further identifications, and by the end of 1963 nine of the small-diameter radio sources, superficially starlike in appearance, had been identified as distant extragalactic objects of this hitherto unknown class. At that time the determi-nation of the redshift of one of these sources—3C 147—had implied a distance of over 5 billion light-years, greater than that of the radio galaxy 3C 295 in Boötes. These strange objects, at first mistaken for stars in the Milky Way, turned out to be distant extragalactic constituents of the universe, generating immense amounts of energy in comparatively small volumes of space. Ini-tially they were known as quasi-stellar radio sources, but soon the acronym *quasars* became generally accepted.

In the years that followed, many more quasars were discov-ered and were recognized to be an important constituent of the universe. Now, more than twenty years after the recognition that we were dealing with a hitherto unknown type of celestial object, about ten thousand extragalactic objects have been identified with radio sources. Some 30 percent of these are quasars, and the majority of the remainder are radio galaxies. These first series of measurements made with the Jodrell telescope may well remain among the most significant ever achieved. Although Henry Palmer has never received any of the acclaim for the discovery of quasars, it was his measurements that led to the search and to their eventual recognition as a new class of object.

Cosmological Issues: Big Bang versus Steady State

The early incentive ultimately leading to the discovery of the quasars came from the Jodrell Bank measurements that revealed the existence of radio sources with angular diameters of less than a few seconds of arc. The Mark I telescope became operational at a critical moment in the development of these researches, and for many years more than a third of its operational time continued to be devoted to this work. By 1961 the diameters of 384 sources had been measured by means of the Mark I linked to the remote aerial on a Lincolnshire airfield. Five of these sources were still unresolved, which meant that their angular diameters must be less than 0.8 seconds of arc. At this stage it was expected that the angular-diameter measurements would lead us to a decisive conclusion about the major cosmological issues.

At that time scientists were divided in an often bitter dispute about the origin of the universe. One group argued in favor of continuous creation or the steady state theory, based on the concept that the universe possessed a high degree of uniformity in space and time. The expansion of the universe was accepted as axiomatic, but it was argued that its unchanging nature was forever established because of the continuous creation of hydrogen atoms. The rate of creation was such that although galaxies would eventually disappear over our observational horizon, others would be created to take their place. This *perfect cosmological principle* implied that the universe had existed for an infinite time in the past and would exist for an infinite future.

The other group of scientists adhered to the more conventional view of an evolutionary universe. In this case the expansion we observe today implies that in a past epoch the matter of the universe was highly condensed. The nature of the initial condensate and the epoch of the initial moments of the expansion were not well defined. The ideas about the nature of the early state varied from Georges Lemaître's concept of the primeval atom to George Gamow's notion of a hot, explosive beginning, according to which the elements were created in the first seconds of time. The look-back time to the initial state was established then, as it is today, primarily from the measured rate of recession of the galaxies. The time scale is uncertain to at least a factor of two—it could be ten or twenty billion years.

Hoyle, one of the chief protagonists for the steady state theory, had pointed out that the angular-diameter measurements gave the best immediate hope of subjecting cosmological theory to observational test. He had drawn attention to the peculiar difference between the apparent angular diameter of a source of a given size according to the two theories. As the distance of a source of a given size increases, the apparent angular diameter decreases, and according to the steady state theory the apparent diameter decreases continuously. However, one of the strange predictions of general relativity is that this would not happen in an evolutionary universe. With increasing distance the apparent diameter at first decreases, as in the steady state theory, but reaches a minimum and then increases as the distance continues to increase. This diameter at which the minimum occurs is calculable and, given the cosmological parameters then available, lay within the scope of the Jodrell angular-diameter measurements.

Although many hundreds of diameter measurements were made at Jodrell with this vital cosmological issue as the prime motive, their interpretation remained uncertain. The discovery of the uniformly distributed microwave background radiation in 1965, and its interpretation as a relic radiation from the hot initial condition in the early state of the universe, soon led to the abandonment of the theories of the steady state, in favor of the evolutionary, or hot big bang, concept.

The Nature of the Radio Galaxies and Quasars

The quasars provided a further incentive for the pursuit of the angular-diameter measurements. Once more it seemed that the radio sources of smallest diameter could be used as the targets for the optical identification of even more-distant quasars. To this was added the mystery of the great energies produced in the quasars. Even with the upper limits of diameters now set to the unresolved sources, it seemed that these distant quasars must be generating energy equivalent to millions of suns in a small volume of space.

At the moment when these strange features of the quasars were emerging, I went to the Soviet Union as the guest of the Academy of Sciences. This generous invitation was arranged by the academy to thank me for all that we at Jodrell had done to help Soviet scientists track their space probes. It was the summer of 1963, and at the Crimean Astrophysical Observatory a summer school for young Soviet astronomers was in progress. The brilliant Soviet astrophysicist I. S. Shklovsky was there, and after one of my lectures we had a long discussion during which he displayed great excitement about the physical nature of the quasars. He was convinced that the few that were unresolved were immensely distant objects in the universe, and he urged that we should extend our interferometers until we could discover their angular extent. To the purely cosmological case for improved resolving power Shklovsky had added the physical argument that entirely unknown processes of energy production might be operating in these remote regions of the universe.

I returned to Jodrell with the determination to set closer limits to the angular sizes of these unresolved quasars. There were many problems, not all of them technical. Palmer and his group had spent an arduous winter in establishing a remote base at Pocklington, in Yorkshire, and wanted to continue with that site. But this would not improve our knowledge of the five sources we already knew to be less than 0.3 seconds of arc in angular extent. Eventually we found the solution with an arrangement to use an 82-foot radio telescope that had been built on the runway of the airfield I had used during the war at Defford, in Worcestershire. The airfield was no longer operational, and this radio telescope was one of a pair that had been built for measurements of interest to the defense department. The distance from Jodrell was almost the same as that to Pocklington, but the Defford telescope was larger than our mobile aerial and could be used on a wavelength of 20 centimeters instead of the 70 centimeters we had so far been using. This immediately improved our resolving power by a factor of three and a half. By May 1965 the new arrangement was working, and the five sources previously measured were found to have an angular extent of 0.1 second of arc or less. These were remarkable results. Clearly we were concerned with physical processes in these quasars, emitting an energy equivalent to millions of suns but contained within a spatial volume that was minute in comparison with galactic dimensions. We continued to reduce the wavelength of the Defford-Jodrell system, and these new measurements on another fifty sources showed that the spatial volumes involved were to be compared only with volumes of space containing a few of the nearer stars in the solar neighborhood.

For more than a decade we had led the international field in the search for higher and higher resolving powers. When the construction of the Mark I radio telescope began, the achievement of resolving powers in the radio spectrum equaling that of the unaided eye was an unfulfilled dream. Now, in 1966, the telescope that had such a turbulent history had been instrumental in the attainment of resolving powers in the radio spectrum thirty thousand times superior to those available in the early 1950s—not only better than the human eye but ten times superior to the resolution of the largest optical telescope on earth, operated under the best seeing conditions.

But we had reached our climax. In order to extend the interferometer baseline even further we had proposed a new technical solution during my visit to the Soviet Union in 1963. This proposal was to establish an interferometer system between Jodrell Bank and the large Soviet telescope on the Black Sea coast. To extend the present technique would have required fifty radio links, and this was scarcely feasible either technically or economically. We therefore made a new proposal. This was to make tape recordings separately at the site of each telescope and subsequently bring the tapes together and correlate the signals. The recommendation was to develop a phase correlation system using high-quality tape recorders, wide bandwidths, and precise timing and synchronization systems. Unknown to us, astronomers in the

United States and Canada had been stimulated by the Jodrell measurements to seek a similar technical solution to the problem of multimillion wavelength baselines, and in this work we had neither the manpower nor the financial resources to compete in their time scales. The protracted and curious history of our attempts to carry through this program with the Soviets has been described in chapter 6 of my book *Out of the Zenith*.

Eventually, in March 1969, our system operated successfully, but this was nearly two years after the Canadians and the Americans had succeeded with similar independent tape recording systems. In November of that year we first collaborated with the Canadians and Americans in establishing transatlantic interferometer baselines over a distance of 6,400 kilometers—a baseline of 13 million wavelengths giving a resolving power of 0.003 seconds of arc.

It was ironic that the system we had first proposed as a collaborative venture with the Soviet Union in 1963 eventually involved us with the Canadians and Americans and not with the Soviets. In the interim Shklovsky had at last been allowed to travel outside the Soviet Union and had visited the National Radio Observatory at Green Bank, in West Virginia. Whether it was his transmission of the ideas in our memorandum to the Americans that started them and the Canadians on their independent tape recording system, I do not know. Perhaps some future historian of international scientific relations will be able to establish this interesting point. At least some of our own students who migrated to West Virginia during those years had the impression that this was so.

Even with these multimillion-wavelength interferometer baselines many quasars and radio galaxies still could not be resolved, and this remains the situation today.

Pulsars

The existence of the radio waves from space was discovered accidentally by Karl Jansky in 1931, and the subsequent history of the subject has been full of accidental discoveries. No other story has been as sensational as that of the discovery of pulsars.

On the afternoon of 21 February 1968 a meeting of the Science Research Council* was about to begin in London when Fred Hoyle came to a vacant chair on my left. He had just returned from the United States. At that time the work on the distant quasars was almost monopolizing the attention of astronomers—particularly the measurement of their redshifts using the large American optical telescopes. That was on my mind when I asked him if there was any news. "Not much about quasars, but last night at a colloquium in Cambridge, Tony Hewish announced that he had discovered some radio sources which emitted in pulses with intervals of about a second." This seemed so unlikely that I made the obvious response that it must be some odd form of interference. To this Hoyle replied that the evidence for an extraterrestrial origin seemed convincing, that Hewish had been investigating these pulsed emissions for months in great secrecy, that four had been discovered, and that all their characteristics pointed conclusively to an origin in the Milky Way.

The following day when I returned to Jodrell I was astonished to realize that, although our contacts with Cambridge were close, no one had any foreknowledge of this discovery. Indeed, it soon became evident that Hewish and his small group had for several months achieved a screen of security and secrecy about this work that was, in itself, remarkable in an age of instant communication between astronomers. Eventually Hewish was awarded the Nobel Prize for this discovery, but many astronomers still maintain that this honor should have been shared with Jocelyn Bell, who first discovered the phenomenon of the pulsations by careful scrutiny of extensive records.

The history of the discovery of the pulsating radio sources (soon known as pulsars) is a strange example of serendipity. In March 1965 the Science Research Council awarded Hewish a small research grant (£37,000).† It was a minor award for the construction of a small aerial so that he could study the phenomenon of

*In 1981 the name of this organization was changed to the Science and Engineering Research Council.

†Hewish's application was received in the transition period from the DSIR to the SRC. The award of the grant was agreed at the penultimate meeting of the astronomy subcommittee of the DSIR in March 1965 and was subsequently processed through the SRC.

interplanetary scintillation—the fluctuations in strength of the signals from a distant radio source that arise when the source is viewed through the solar corona. Routine recording with this new equipment began in July 1967. The record of the signals received by this equipment appeared in the form of a pen recording, and the paper charts accumulated at the rate of about four hundred feet per week. One of Hewish's research students, Jocelyn Bell, had the task of making that part of the analysis of these records concerned with the celestial coordinates of any scintillating component of the received signal. On these records there appeared a scintillating source in transit near midnight, a time when the interplanetary scintillation effect becomes very small. By the end of August she had plotted this source on several occasions. Although conspicuous on the records it was sporadic. After various tests Hewish had by the end of September concluded that the radio source responsible for this scintillating effect on the records must lie outside the Solar System. After more investigations there seemed no doubt that this pulsating source was located somewhere among the stars of the Milky Way, and the Cambridge group concluded that the most likely candidates were either pulsating white dwarfs or neutron stars.

The records that had yielded this remarkable information were naturally searched for other pulsating sources as soon as the galactic nature of the first was established. Jocelyn Bell reexamined the three miles of charts and within a few weeks found three other pulsating sources in widely differing parts of the sky. At this stage, when Hewish and his colleagues were certain that the pulsating signals arose from some kind of stellar objects in the Galaxy, they published the information. When this communication appeared in *Nature* in mid-February, astronomers everywhere were amazed. Continuity was the order of the day in astronomical observations. True, sporadic outbursts in the optical and the radio wave bands were well known—those of supernovae and solar flares, for example—but these were emissions of continuous radiation. The idea that celestial sources could produce pulses of energy required a fundamental reorientation of thought. It is perhaps scarcely surprising that many individuals, including a number of scientists, thought in terms of an extraterrestrial civilization as the origin of these pulses.

The amazement at the publication of the results was accompanied by a good deal of annoyance that the Cambridge scientists had behaved with such secrecy about their discovery. However, it must be remembered that Hewish and his colleagues were faced with a quite extraordinary situation. When the signals were first noticed, in the summer of 1967, the natural reaction was that they were some form of the terrestrial interference that is the plague of every radio astronomer. Slowly it became clear that the origin of the signals was somewhere in the sky outside the earth's environment. Not until the end of November was the pulsating nature of the signals established, and it must have been hard even then to believe that they had a natural origin. Furthermore, other evidence showed that the pulses came from an object whose dimensions were about the same as those of a planet and that the pulses were not constant in strength but variable. It was then necessary to make observations to find out whether this variability was indeed a code and whether the signals had an intelligent extraterrestrial origin. Hewish and his students must have found it difficult to keep this unfolding drama to themselves. They did so until they were able to produce the incontrovertible scientific evidence that the objects were among the stars, that there was no coding, and that the energies involved were those to be associated with natural phenomena.

At the time of my conversation with Fred Hoyle at the meeting of the Science Research Council, when I first heard about the Cambridge discovery, the Jodrell Bank telescope was being employed on a flare star program of my own during the night, and during the daytime the same equipment was used by another group to study the interplanetary scintillation phenomenon. The telescope seemed to be ideally equipped to study the pulsating sources. Whereas the Cambridge telescope could observe the pulsating source for only about four minutes per day, the Jodrell telescope could follow it continuously as long as it was above the horizon, and consequently information about the characteristics of these newly discovered radio sources accumulated rapidly. After only two weeks' work these new data were published—the second of the great flood of communications soon to appear in *Nature* from many scientists all over the world.

At Jodrell nearly a year elapsed before pulsar observations began to take a more normal share of the observing time of the telescopes. As I write now, twenty years after their discovery, the research on the pulsars still occupies a quarter of the operational time of the telescope. Ninety have been discovered with the instrument, about a third of those observable in the Northern Hemisphere. Altogether about 450 pulsars are now in the catalogs. They are neutron stars—the highly condensed remnants of a normal star that has ended its life in a supernova explosion. Stars like the sun collapse into white dwarfs when all their energy-producing processes are exhausted, but a star more than about 1.4 times the mass of the sun will end its life in the cataclysmic event of the supernova—the most famous of which is the Crab nebula, the supernova event recorded by the Chinese astronomers in A.D. 1054. The condensed nucleus of the Crab is observed as a pulsar, but although the pulsar-supernova relationship seems secure, only three visible remnants are known to coincide with the pulsar. The visible remnant of the supernova expands at a high rate and becomes invisible in a time that is short compared with that of the remnant neutron star observed as a pulsar. The theory of the radio emission from the neutron star, which makes it visible as a pulsating radio source, is not yet completely understood, although it is generally agreed that the emission is associated with the rapid rotation of the neutron star and with its very high magnetic field. As far as Jodrell is concerned, the discovery of the pulsars was a most significant event. The telescope, with its high gain and steerability, was a unique instrument for the study of the phenomenon and the search for undiscovered pulsars. Unexpectedly we had been presented with a compelling incentive to ensure the life of the instrument beyond any reasonable expectation.

The Gravitational Lens

Although the study of pulsars has been significant in the work of the telescope since 1968, the solid core of the researches has concerned the nature of the distant radio sources—quasars, radio galaxies, and other extragalactic objects. There have been surprises in this work, but perhaps only one that could parallel the discovery of the pulsars.

On 30 March 1979 the Jodrell extension telephone in our home rang at a time near dawn that I normally associated with some emergency. It was the duty controller to say that Dr. Walsh wished to speak to me urgently. Dennis Walsh, a member of the Jodrell staff, was in the United States at the Kitt Peak National Observatory, in Arizona. He had been given a few nights' observing time with the large telescope there for an attempt to make optical identifications of some of the radio sources he was investigating at Jodrell, and I could not imagine what catastrophe had befallen him. However, it was not a voice of trouble but one of excitement that came over the telephone. He believed that he and two colleagues, R. F. Carswell from Cambridge and R. J. Weymann from the University of Arizona, had just made an extraordinary discovery. He had expected to find a single remote quasar in the position of one of his radio sources. But there were two very close together with apparently identical characteristics and with the same redshift, implying that they were at the same distance. The chance that these could be two separate objects close together in the heavens at that distance of several billion light-years was so remote that the idea could be rejected, and Walsh and his two companions had concluded that they had discovered the phenomenon of a gravitational lens.

In 1972–73 Dennis Walsh and two other members of the Jodrell staff (Ian Browne and Ted Daintree), together with two research students, used the telescope to make a detailed survey of the radio sources in a limited area of the sky. The idea was to obtain a complete sample up to the sensitivity limit of the instrument and then search the Palomar Sky Survey plates for optical identification. About 780 sources were cataloged. Various categories of galaxies and other visible objects were readily identified with 315 of the radio sources, but 217 (or 40 percent of the radio sources for which identification might have been expected) lay in positions on the sky survey where no photographic objects could be found. This raised a critical question. Would the optical search for objects too faint to be on the sky survey plates yield identification of the radio objects with more-distant galaxies or quasars?

This could be settled only by using one of the large optical telescopes in the Northern Hemisphere to search the sky in the position of the unidentified radio sources to fainter magnitudes than those recorded in the Palomar Sky Survey.

The Zelenchukskaya Diversion

Even a few nights' observing time on the large optical telescopes available at that period was difficult to obtain. I then had what appeared to be the excellent idea of arranging for Walsh and Browne to use the new Soviet 6-meter optical telescope. This was the largest in the world, but little was known about it in the West. In October 1976 my wife and I received an invitation to visit the site of the telescope. There was no particular reason for this invitation since I was not an optical astronomer, but at that time our relations with the Academy of Sciences were very good and we had enjoyed extending hospitality to a number of high-ranking Soviet scientists in our home as well as at Jodrell. I had often asked about the progress with this instrument and with a large radio telescope being built in the same area of the North Caucasus, and this invitation was a natural response. At least in those days a visit to the telescope was not an easy matter, and even as guests of the academy we found ourselves on a perilous enterprise.

After a two-hour flight from Moscow to the airport of Mineralnye Vody we reached the remote observatory by a 150-mile car trip, the last 30 miles of which were up the side of a mountain. The visit went well until heavy snow fell during the night before our departure. The precipitous mountain road had become nearly, but not quite, impassable. The 150 miles to the airport seemed more like a thousand; since the journey was by zigzags across the icy surface, we may well have traversed that distance.

Our arrival at the airport several hours behind schedule was of no consequence, since it was both ice- and fogbound. No flights were departing, and we could not imagine how the vast crowd occupying every inch of the floor of the airport building would ever leave. We never discovered, because by one of those strange near miracles that seem to occur only in the Soviet Union a charming English-speaking hostess appeared in the early morning hours. "If you will walk with me across the ice, there is a small plane load of Armenian footballers who insist on leaving, and you can join them." We were hauled into the plane from the rear and arrived at our friend's house in Burakan in time to breakfast on the dinner that had awaited us the previous evening.

Undeterred by this experience, I urged Walsh and Browne to allow me to use the promise of collaboration to facilitate their use of this, the most powerful optical telescope in existence, so that they could identify their blank field sources. Eventually there were three attempts at collaboration, by Walsh and Browne in 1978–79, but each was thwarted by either bad weather or technical problems.

The Kitt Peak Observations

Meanwhile Walsh had successfully asked for an allocation of a few nights' observing on the 2.1-meter telescope at the Kitt Peak National Observatory. He used the Kitt Peak telescope, with its sophisticated recording equipment, to study two of the already identified objects in the Jodrell survey. Although their identification in the Palomar Sky Survey as two distant quasars seemed secure, one peculiarity needed further study. The two objects were of the same magnitude but were very close together in the sky, being separated by only 6 seconds of arc. On the night of 29–30 March, Walsh, working with Carswell and Weymann, discovered that these two quasars were really identical. They had identical spectra and the same redshift, indicating they were at the same distance. In fact, the various characteristics left no doubt that the two objects were associated. The chance that two entirely separate identical objects could be so close together in the universe at that distance was negligibly small. The account of this work published in *Nature* on 31 May 1979 assembled evidence that the two quasars were not separate objects but two images of the same object formed by a gravitational lens created by a massive intervening object close to the line of sight. The intervening object—a large galaxy—much closer to earth than the quasar, was later detected by a group using the 200-inch Palomar telescope.

The deflection of a ray of light passing close to a massive body is predicted by gravitational theory. The famous eclipse expedition of 1919 verified that the light from a star passing close to the sun was deflected in accordance with the predictions of Einstein's general theory of relativity and that the deflection was about twice as great as that to be expected on the basis of Newtonian theory. The influences of these predictions on the observations of the distant universe did not appear to be of significance until the extent of the extragalactic universe became apparent in the 1930s. Shortly thereafter various scientists drew attention to the possible modification of the light from a distant galaxy if it traversed another massive galaxy in its passage to an observer on earth.

This discovery by Walsh and his two colleagues was quite accidental; it was not based on any attempt to find observational verification of these theories. It is another most curious illustration of the unpredictable nature of progress in science.

The Flare Stars

The study of the flare stars does not tend to be regarded by astronomers as having the kind of fundamental significance to our understanding of the universe as the work on radio galaxies or quasars, for example. I write about them here because they became the main subject of my own research with the telescope.

When the telescope became operational, many people in my position would have assembled a group, I imagine, or joined one of the existing groups, to pursue the investigation of the cosmic radio emissions already in progress at Jodrell. I did not do this. With this new and powerful instrument it seemed to me that fresh paths should be explored beyond the limits of conventional wisdom. It is hard to translate this type of idealism into practice without wasting time on hopeless enterprises. There must be a rational and scientific background to any such quest, and circumstances soon directed my attention to an ideal problem.

In the summer of 1958 I attended a symposium at the Cité

Universitaire in Paris. Evry Schatzman, then of the Institut d'Astrophysique and Faculté des Sciences de Paris, gave a brief and highly speculative talk entitled "On the Possibility of Observing Radio Emission from Flare Stars." At that time I had no idea what type of star Schatzman was referring to, and I doubt if many others present were in a better position. In fact, this class of star had been recognized only ten years previously. Willem J. Luyten of the University of Minnesota had discovered that one of the closest stars to earth—UV Ceti—had unusual properties. On 7 December 1948 he observed that the star had abruptly increased in brightness by two magnitudes for a short time. He wrote, "The sudden flaring-up appears to be a rare phenomenon whose ultimate solution should aid in the solution of the more general problem of the source of stellar energy."

Ten years later, when I began work on these stars, some two dozen were known—all were red dwarf stars of small mass compared with the sun's and of low temperature (less than 3000 degrees Kelvin compared with the 6000 degrees of the sun). The sudden increase in brightness of these stars of half a magnitude or so occurred every few hours, and there were occasional instances of far more dramatic increases of five or six magnitudes. The sun exhibits sunspots and flares, and the associated bursts of radio emission are powerful and easy to observe with simple equipment. But the nearest of the red dwarfs is 300,000 times farther away from earth than the sun is. If flares with the same intensity occurred on the red dwarf and on the sun, the one on the red dwarf would appear some hundred billion times weaker to an observer on earth.

Although the Jodrell telescope was far more sensitive than any other radio telescope available on earth, there would have been no possibility of compensating for this dilution in the strength of the signal. There were, however, some less discouraging aspects to the red dwarf flares. The most significant appeared to be that the energies of the frequently occurring red dwarf flares were a hundred to a thousand times greater than those emitted in the most violent solar flares. That, at least, was the scientific justification for the use of the telescope in an attempt to detect the radio emission from these flares. If the telescope automatically followed

302

one of them, there would be, I thought, a reasonable chance of detecting a radio flare about once in every twenty hours of observing time. This was unknown territory, but on 28 September 1958 I started work with the telescope in automatic motion following the star UV Ceti—the star observed by Luyten ten years previously. In those days before computerized data recording, the output of my receiver was fed to a pen-and-ink recorder. A sudden increase in signal, as might be expected from a flare, would have caused the pen to deflect across the chart—but for how long was quite unknown.

After some experience it soon became evident that the type of low-intensity short-duration signal for which I was searching was only too readily produced by many terrestrial sources, so I mounted another aerial at the focus of the telescope, placed so that I had two aerial beams moving across the sky, one on the star and the other on a blank comparison region. Any interference from a terrestrial source should appear in both of these channels while a real increase in signal from a flare would appear only on the star channel. After eighteen months I had observed UV Ceti and four other flare stars for nearly five hundred hours in this way. When all doubtful cases had been excluded, there were thirteen events compatible with bursts of radio emission from the flare star, registering on the star channel but not on the comparison channel.

When I showed the results to Hanbury Brown, he said I could not publish them. "Why not?"—"Because nobody would believe you. Get someone to look at the star with an optical telescope and find out if this kind of record really is associated with the flare." After that setback it was my good fortune that I showed my radio records to Fred Whipple when he visited Jodrell in the summer of 1960. He immediately suggested that the Baker-Nunn cameras of the Smithsonian's satellite tracking network should be used in a collaborative program.

This program began on 28 September 1960, and a year later we had accumulated 727 hours with the radio telescope. There were 166 hours of overlap where both radio and camera records were of good quality. During this period there were twenty-three minor flares of less than one magnitude, and since there was no obvious deflection on the radio charts at these epochs, we integrated them by superimposing the radio records aligned on the epoch of the peak of the photographic event. The result of the integration showed an increase in radio emission over the interval -2 minutes to $+8$ minutes about the epoch of the flare maximum. A comparison with a similar integration of the charts where no flare occurred showed that the chance that the increase was spurious was less than one in one hundred million. At last we had the first conclusive evidence for the association of radio emission with flares on the red dwarf stars.

Our collaboration with the Smithsonian cameras continued, but even with the five cameras we lost considerable time because of poor sky conditions. Fortunately, when I was in the Soviet Union during the summer of 1963, I found many of the Soviet astronomers were greatly interested in the problem of the flare stars, and the Academy of Sciences readily agreed to a joint observing program. By the autumn of 1963 this was tied in simultaneously with the Smithsonian network, and the observations instantly achieved an entirely new character. By December 1969 we had observed for over four thousand hours with the radio telescope, and there were many cases of radio/optical correlations of individual flares.

In 1972 a fortuitous circumstance placed me in contact with L. H. Mavridis of the University of Thessaloniki, who, I discovered, had excellent equipment for recording the flares photoelectrically at the Stephanion Astronomical Station, in the Peloponnese. This extension of the international optical observing network soon yielded a splendid result. On the evening of 11 October 1972 there was a flare of such intensity—greater than 4.5 magnitudes—that the pen of Mavridis's recorder went off scale for a few minutes. The record from the Jodrell telescope was a good one, and it indeed revealed a striking increase in the signal strength; however, the sharp increase occurred eight minutes after the initial surge of the optical flare. At last the truth was beginning to dawn. The optical and the radio events were correlated but were not recorded simultaneously on earth. This record stimulated my colleague Franz Kahn in the theoretical astronomy department in the university to study the physical processes in the star when it flared. He concluded that the radio phase of the flare did not begin until the shock wave of the initial disturbance had traveled through the stellar atmosphere to a region where the gas density was low enough for the radio waves to be propagated. The work had finally yielded an insight into the nature of the star—it must have a corona and a stellar wind.

Although this event was a decisive correlation, some puzzling features of this long series of records left me with an uneasy feeling that some of my colleagues were not convinced and felt that the correlations were by coincidence related to some terrestrial burst of radio interference. I longed for a decisive radio-only technique that would pinpoint the increase in signal strength decisively on the star, and my wish was soon fulfilled.

Henry Palmer and his group had been steadily developing more sophisticated techniques for the interferometric measurement of the angular sizes of the radio galaxies and quasars. We agreed to make a test in December 1977 using one of the remote telescopes linked interferometrically with the Jodrell telescope, in a computerized system that produced records from three areas of sky simultaneously—the flare star, a nearby reference source, and a blank area of sky. This system gave excellent discrimination against terrestrial emissions, and it yielded the precise coordinates of any signal appearing on the star channel. In the early morning hours of 15 December and again on 18 December, this system produced the unambiguous records for which we were searching—two radio flares pinpointed by the system to within a few arc seconds of the flare star, with the blank sky channel clear of signal.

Palmer gave me the graphs of the output from the computer a few days before Christmas. I pinned it on a wall at home among the Christmas cards and decorations. As the children and grandchildren assembled, they were mystified by my exclamation that this was one of the very best Christmas presents I had ever been given. Our account of this work was published in *Nature* in June 1978; it was my penultimate publication on the flare stars. In October 1979 there was a massive international effort, involving ground-based optical, radio, and X-ray observations from a satellite, to observe a flare star in Canis Minoris. On 25 October and again on 27 October, flare events were detected in the optical, radio, and X-ray wave bands. For me, the work I had begun alone at Jodrell twenty-one years earlier ended in that way with a published account in which I was one of thirty authors and which involved seven optical telescopes, seven radio telescopes, and the Einstein X-ray satellite—a typical sequence in the development of contemporary scientific research.

When I retired in 1981, I vaguely thought I might continue with some research on these phenomena, but my attention was diverted elsewhere and this did not happen. When writing about this, thirty years after my own observations began, I looked once more at the literature on the subject and was perhaps not entirely surprised to discover the growth of interest in the red dwarf stars. Even so, the radio observation of the red dwarfs has never excited the astronomical community in the same way that the extragalactic phenomena have. At a colloquium held in 1982 under the auspices of the International Astronomical Union (IAU) in Catania, Italy, and entitled "Activity in Red Dwarf Stars," David Gibson spoke on the flare emission from these stars. In 1976 he had worked with me for a few months as a visitor from America, before returning to do research in Socorro, New Mexico. In his address, referring to my early radio detection of flares, he said,

> One might have thought that these unexpected discoveries would have spurred significant interest in this new field but they did not. Why? Three reasons seem most apparent: 1) The "mainstream" of astronomers in the late 60's and early 70's never fully appreciated these data. Cross-fertilization especially with solar astronomers was virtually nonexistent. 2) The data were obtained at very slow rates; e.g. Lovell reported an average of only one flare on UV Ceti per 35 hours observation. In an era of radio studies of "glamorous" new objects as quasars and pulsars such an investment of time was difficult to justify. 3) Credibility. As Lovell and his colleagues wrote, "(detections) were difficult because of the sporadic and transient nature of the phenomena, and the danger of interference. . . . In several thousand hours of observation there are only a few cases where the existence of individual flares of long duration has been unambiguous."[*]

*D. M. Gibson, "Quiescent and Flaring Radio Emission from dMe Stars," in P. Byrne and M. Rodono, eds., *Red Dwarf Stars* (Dordrecht, Netherlands: Reidel, 1983), pp. 273–86.

He then said that the path toward a resurgence in radio studies was paved in 1976 and 1977 by the decisive records obtained by our interferometric records of the flares. They led to the extension of the work by himself in New Mexico and by other groups in the United States. "Not only was quality data available but complementary studies at X-ray, ultra-violet, and optical wavelengths provided the radio observations a new context within the framework of the new field of stellar activity."

And so this remains a field of research in which almost everything has still to be discovered.

21

Mark IA and MERLIN

In the early 1950s many observers maintained that the telescope we were proposing to build would not last for fifteen years, and few scientists believed it could be a useful research instrument for that length of time. In the event, when the thirtieth-birthday celebrations were held, in 1987, the telescope—then renamed the Lovell telescope—was still operating continuously and in greater demand by the astronomical community than ever before. This is a remarkable longevity for a major scientific instrument, and such a long and productive lifetime was scarcely envisaged when the proposal was made in 1949 to build the telescope.

The scientific longevity stems from the great adaptability of the paraboloid. Compared with a large array of dipoles, for example, it is a simple matter to change the operating wavelength, and in that connection the technical changes made during the construction phase to improve the performance at short wavelengths became of great significance. After the telescope came into use there were important advances in the development of sensitive receiving systems for use at wavelengths shorter than a meter. In the beginning we used vacuum tubes for the receiver, and in the 1960s and 1970s the transition to successive forms of highly sensitive solid-state devices presented no problem. Everywhere this stimulated research into the short-wavelength emission from the universe, and the telescope remained one of the most sensitive instruments in the world for these investigations.

Even so, the telescope would not have remained in such an excellent state had it not been for circumstances that led to significant structural changes during 1970–71, resulting from my ambition to build even larger telescopes. As soon as we were freed from the burden of our debt, in 1960, and encouraged by the success of the telescope, I made plans for a whole series of new radio telescopes. For the next twenty years I became involved in a number of political and financial battles that at times were uncomfortably reminiscent of the story of the Mark I.

During the years of research on the angular diameters, we had succeeded in building two more radio telescopes. One of these, known as the Mark II, was smaller than the Mark I but had a paraboloidal surface of greater accuracy. The Mark II was built on the site of the 218-foot transit telescope and came into use in 1964. A telescope of similar size and shape, the Mark III, was a less accurate and more flimsy structure. It was designed specifically as a telescope to be used in association with the Mark I as an interferometer for the angular-diameter measurements. Built in 1966 on farmland 24 kilometers south of Jodrell, it is still firmly integrated into the network of Jodrell telescopes.

These Mark II and III telescopes were subsidiary to my major ambition to build a telescope of immense size. A rough sketch of this proposal made with Husband in 1960 reveals the concept of an elliptical paraboloid with a minor axis of 500 feet and a major axis of 1,500–15,000 feet. This became identified as the Mark IV and caused so much consternation in the grant-giving bodies that a high-level committee was established to reconsider the entire problem of developments in radio astronomy in the United Kingdom. During 1963, when the work of this committee destroyed the vision of this telescope, we began serious design studies on a more conventional instrument—that is, a telescope of the same steerable type as the Mark I but with an aperture of 400 feet—designated the Mark V.

The tortuous sequence of events surrounding my efforts to build this Mark V telescope is described in my book *The Jodrell Bank Telescopes.** Some years earlier, at the request of the DSIR, Ryle and

*Oxford: Oxford University Press, 1985.

I had produced a paper on the future of radio astronomy in Great Britain. We both planned new telescopes of different types for Cambridge and Jodrell, and a special committee under the chairmanship of Lord Fleck (then treasurer of the Royal Society) had made appropriate recommendations. In the spring of 1965, when the new Science Research Council (SRC) first met,* I was asking for a sum of about £4 million to build a telescope far larger (400-foot aperture) than the 250-foot-aperture telescope at Jodrell. If I had been merely an applicant, this instrument would, I believe, have been constructed. However, I was the chairman of the Astronomy Space and Radio Board of the council, charged with exercising a scientific judgment among projects competing within a limited budget. During those five years I voluntarily offered to postpone the Jodrell project, initially in favor of the Anglo-Australian optical telescope at Siding Spring, in New South Wales,† and in December 1967 in favor of Ryle's aperture synthesis radio telescope in Cambridge. These two instruments were built, and both achieved great success. No one could have foreseen that these successive postponements would eventually kill my own project for the major steerable radio telescope. The high inflation and complex political movements of those years eventually led to the demise of the project, which, even after a reduction in size to a 375-foot aperture, had led to estimates for the construction of between £20 million and £30 million when we finally abandoned the scheme in 1974.

From the debris of those confused years two items of positive value emerged for Jodrell. The December 1967 agreement to postpone the large telescope in favor of Ryle's aperture synthesis system was made on the condition that modification and strengthening of the Jodrell telescope would be given high priority. These modifications, made in the years 1969–71, gave us a superior instrument, free of many of the strains and stresses of the original telescope.

*In 1965 the Science Research Council was formed to take over from the Department of Scientific and Industrial Research (DSIR) the allocation of government grants to universities and other institutions for scientific research in the physical sciences.

†I have detailed the history of this phase of the Anglo-Australian telescope project in *Quarterly Journal of the Royal Astronomical Society* 26 (1985):393–455.

The Mark IA Telescope

The stimulus for the bargain I made with the SRC at the 1967 Christmastide arose from a desire to improve the surface accuracy of the Mark I telescope so that it would operate with greater efficiency on shorter wavelengths and also from Husband's warnings about the stresses and strains that were developing in the instrument. The major changes were the building of a massive load-bearing circular girder system that would replace the slender stabilizing girder of the Mark I design and the superimposition of a new solid-sheet paraboloid above the existing bowl. It was also necessary to repair the foundations of the existing railway track and to construct a new inner track to reduce the load on the azimuth track and trunnion bearings. This relatively simple alteration was completed before the end of 1969, without serious interruption to our use of the telescope. The construction and erection of the new load-bearing wheels and a new paraboloid surface was quite a different matter. From August 1970 to November 1971 the telescope was wholly in the hands of the construction engineers.

From the critical decision taken in the office of the chairman of the SRC in December 1967 to the signature of the hand-over documents, the conversion of the Mark I to the Mark IA took six and a half years—as long as the construction time for the Mark I. The cost of the conversion, £664,000, was numerically greater than the cost of the original instrument. During this long period, and especially when we were without use of the telescope in 1970 and 1971, I had repeatedly complained to Husband about the delays. To my surprise he replied that it was the most difficult engineering job he had ever tackled. As the costs of the work soared and as the likelihood of attaining the Mark V project became smaller and smaller, Rainford, with great foresight, remarked that the decision to do the conversion was the wisest one we had ever made.

The MERLIN System

The final decision to abandon plans to build the large telescope was also compensated for by grants for the alternative multitelescope system, now known as MERLIN (multi-element radio-linked interferometer),* which since 1981 has produced the splendid high-definition radio maps of the distant galaxies and quasars.

The idea of MERLIN arose from a suggestion made to us by Henry Palmer late in 1973, when the Mark V telescope project seemed on the point of rejection. In principle he proposed that we should add more telescopes to the existing Jodrell-Defford interferometer so sited that, when the signals from all the telescopes were combined, we would have a radio source map showing detail of less than a second of arc. It soon became clear that by far the most economical means of proceeding would be to purchase telescopes from an American source. The firm E-Systems, of Dallas, was in the process of manufacturing twenty-eight telescopes of 82-foot diameter for the American VLA (very large array) being built in New Mexico. Eventually E-Systems supplied and installed three of these telescopes in our network with such efficiency and speed that they were available for use long before our own complex electronic developments were ready. As 1980 was drawing to a close the complete system came into operation. Six telescopes operating simultaneously on the same area of sky began to produce maps of a quality and definition that in subsequent years formed an important and international aspect of the researches at Jodrell.†

We are too close to these events to judge the correctness of the decisions. The instant reaction is that the sacrifice of the 400-foot telescope in favor of the alternatives was the correct procedure. That is the easy answer because all of these alternatives have led to highly successful research instruments. Until someone, someday, constructs the massive telescope I wanted, the judgment

*The stories of how the Mark I became the Mark IA and of MERLIN are given in my book *The Jodrell Bank Telescopes* and in the last chapters of *The Voice of the Universe.*

†The further extension of the MERLIN network is expected to be complete in 1990 by the addition of a 32-meter telescope on the radio astronomy site at Cambridge.

cannot be final. Only then will we know what could have been done in these years.

A Survival Test for the Mark IA

I have said that the survival of the Mark IA telescope as a front-line instrument for research is a phenomenon of our age. Scientifically a twenty-four-hour schedule has never been able to satisfy those who want to use it. As an engineering structure it has posed few problems. Husband wisely built in a redundancy that has undoubtedly saved the telescope from destruction in wind forces that on one or two occasions have significantly exceeded the recommended safety factors. Nevertheless, on 2 January 1976 I was given a sharp reminder that, like a ship at sea, the telescope was in constant danger from the elemental forces of nature. When I left Jodrell that evening, the telescope was already locked in the zenith because a high wind had been blowing all day. On my homeward journey the few miles of lanes through which I drove were scattered with twigs and small branches from the trees, and I was thankful to park the car and reach the comfort of the fireside. About an hour later a great tree crashed to the ground near the house, and as I reached for the telephone that would connect me with the Jodrell control room, the controller phoned to urge me to return to Jodrell. The huge wheel girders had been shifted more than seven inches across the upthrust units by the force of the wind. Another inch and the entire 1,500-ton weight of the bowl structure would fall on the trunnion bearings, and the whole elevation structure of the telescope would probably collapse.

With the greatest difficulty I fought against the wind to reach my car, only thirty yards from the house. The village seemed deserted, and not a single vehicle or person was to be seen on my route to Jodrell. I reached the control room to find that the mean wind speed was over seventy miles per hour and still rising and that the pen of the barometer was falling as though over a steep cliff. By this time only half an inch of the wheel girder on the bogies of the upthrust units was relieving the trunnion bearings of an insupportable load. There was only one possible solution—

to rotate the telescope through 180 degrees so that the great force of the hurricane would shift the wheel girders back onto the upthrust units. In that way the telescope would become athwart the wind and thus move through a condition for which the wind rakers and outer bogies had never been designed or subjected. With the wind now over ninety miles per hour and reaching hurricane force 12, we switched to our diesel generators, in case the main supply should fail at a critical stage, and set the telescope in slow azimuth motion. After some 45 degrees of rotation I saw with immense relief the sudden jump of the wheel girders as they slid over the upthrust units to their normal position.

By the time we completed this move, the worst of the storm was over and the telescope was safe. By daybreak there was still a gale force 8–9, but we were able to inspect the vital parts of the telescope structure. There was no obvious damage. Clearly the telescope had been built to survive. Husband was far away from the center of this hurricane and had no fear on that night for the safety of the telescope. When he eventually analyzed the problem, he concluded that the great force of the wind on the upturned bowl had actually tilted the towers from the vertical and thereby shifted the whole bowl transversely so that the load-bearing wheel girders had nearly been removed from the upthrust units fixed on the diametral girder. In the Mark I the "bicycle wheel" was merely a stabilizing device. It carried no load, and such a shift would have been of no consequence. Now in the Mark IA modification a significant part of the 1,500-ton load of the bowl structure was taken to the inner railway track through the wheel girders. Fortunately the solution was relatively simple—the addition of steel bracing girders to the two towers.

Two Million Visitors

In the early 1960s the constant publicity because of the association of the telescope with the Soviet and American space projects led to large numbers of requests from individuals and groups who wished to visit Jodrell. We were a department of the university charged with a duty to educate as well as to carry out research, but we soon found it impossible to receive even the most deserving groups of visitors. Eventually I suggested to the vice-chancellor that we make some provision for satisfying this legitimate public interest, along the lines that other observatories, particularly in America, had found necessary. He doubted whether the demand would justify the expense and suggested that I might test public reaction by erecting a marquee and holding two weeks of open days during which I would give explanatory lectures. In the summer of 1964 we made this experiment, charging a half-crown entrance fee to cover the cost of the marquee. On the first Sunday afternoon the queues of cars waiting to gain access extended for two miles, and a total of 35,000 people came in those two weeks. Even so, the university officials were not convinced that there was sufficient public interest to justify any permanent arrangement. The experiment was therefore repeated during the summer of 1965—with a similar result.

After these experiences of 1964 and 1965 the university agreed that we should erect a building that would serve the needs of visitors on a regular basis. This building, originally known as the Concourse Building, was officially opened on 3 May 1966. In 1971 it was extended to include a planetarium, and in 1972 more than thirty acres of adjoining land were planted as an arboretum under the auspices of the university botanical department.

The exhibits in this center have constantly been revised. Originally intended to concentrate on explanations of the research work at Jodrell, they have over the years been expanded in scope, to include interactive exhibits necessitating participation of the visitors. Now renamed the Science Centre, the complex is a great success and attracts more than 100,000 visitors every year, nearly half of whom are schoolchildren. The millionth visitor arrived in March 1978, and in 1989 I welcomed the two millionth. In an age when countless complaints are made about the public relations of science and scientists, we feel that the Science Centre has satisfied a public demand without interfering with the scientific work at Jodrell. It was the sight of a huge electric arc that determined my career as a scientist, and I would be happy if the close-up view of the telescope has done for a few of the million young visitors so far what the electric arc did for me.

22

Diversions from the Workbench: Affairs of Defense

Throughout the years of progress at Jodrell described in the preceding chapters, I became seriously diverted from the task of astronomical research. Substantial diversions from the work for which they were trained have become inevitable for many scientists in the Western world. On any working day when scientists might reasonably be expected to be concentrating their thoughts and efforts on research, a significant percentage will be involved in a maze of committees dealing in some way with the administration or finance of research. To a large extent the rapid increase in these diversions is the result of the vast growth in the complexity and cost of research during the last few decades and of the state's need to protect the expenditure of the considerable sums of money now involved. The financial problem is exacerbated by the urge of scientists to pursue their own specialty. Since no society can now afford to be preeminent in every discipline, this leads to infighting between scientists and to their search for international collaboration, further diverting scientific effort to organizational and political problems.

In these matters, I am sorry to say, I write from experience, some of which has not been without bitterness. The decisions lie with the individual, but the dividing line between an isolated selfishness and service to the scientific community and the state is ill defined and hard to draw. As for myself, I do not think now that I was able to maintain an appropriate balance. However, it is not yet possible to pass judgment on those years in the context of history, and my aim is simply to describe what happened in my own life.

The Aeronautical Research Council

After six years of war, as I described in earlier chapters, my one desire was to return to an academic life in the university. For a few years I did so, and those years of concentration on research in the late 1940s, when Jodrell was being created and I was learning about the universe, now seem to stand in isolation from the years that followed. At least in the minds of those charged with the defense of the realm, a new line of potential conflict was being drawn across the map of Europe. Toward the end of 1952, just over seven years after I had passed through the gates of the wartime establishment for the last time, the first approaches were made to me on the grounds that my presence on a committee of the Aeronautical Research Council was desirable. It was one of those quid pro quo approaches that, at least in Britain, seem an inherent part of civilized life. The backing of Sir Harold Spencer Jones, the Astronomer Royal, for the telescope had been of cardinal importance. Now, it appeared, he needed my help on the Aeronautical Research Council's Air Navigation Committee, of which he was the chairman. The terms of reference included assessment of the relative advantages of alternative bombing methods. I could not plead ignorance of the subject or deny that my postwar technical developments might have a bearing on these issues.

I attended my first meeting of this committee in February 1953 and was soon immersed again in issues I had left behind in 1945. Before the year had ended, I was leading a double existence, producing one paper with the argument that the new postwar air-borne radars were "fundamentally useless for bombing purposes" and that the principles had been neglected in favor of aircraft design and one paper "on the use of extraterrestrial radio waves for astronavigation." This soon led to a proposal that the Ministry of Supply help with the finance of the telescope so that the essential measurements could be made on the radio stars with a view to producing an accurate navigational aid of use under overcast conditions. In the earlier chapters on the building of the telescope, I already wrote about the maze into which that led me.

On once again reading the files of those years, I now see clearly how perilous my career became. Blackett had left Manchester for Imperial College, London, and, like the father figure he was, sounded the alarm bells in a letter to me in March 1954: "For God's sake don't take on too much! If you cracked up, I suspect that the whole project would be abandoned. . . . My advice would be to reduce to a minimum all diversions and concentrate (a) on getting the mirror [telescope] going and (b) on deepening your own scientific work. Forgive the lecture!"

Blackett's no doubt sound advice was too late. The idea that the ministry should help finance the telescope for researches into the navigational problem fell on stony ground, but I was asked to serve also on the Air Warfare Committee. My activities on these two committees of the Aeronautical Research Council led to a further request in February 1955 that I serve on the council itself as an independent member. My protestation of concern that I could not possibly attend all the meetings in London of the two committees and the council had no effect, and from June of 1955 I began to serve on the council and as vice-chairman of the Air Navigation Committee. At the height of the crisis with the building of the telescope these diversions might seem, in retrospect, to have been distractions that might well have been avoided. At the time I clearly thought otherwise. Not only were the affairs of the telescope involved, but my interest in matters of defense had been thoroughly reawakened, and I began to feel that it was a fulfillment of a duty.

My three-year membership of the council ended in March 1958. For another two years I remained a member of the Air

Warfare Committee but attended few meetings. Although Blackett rightly warned me of the dangers in his letter of March 1954, it is perhaps as well that I had distractions external to the worries of the building of the telescope during those years. At least it is undeniable that the telescope would not be as it is today if I had rejected the approach from Spencer Jones for help on the Air Navigation Committee early in 1953. The urge to improve the accuracy of the paraboloid to make it useful for defense activities would not have existed, and its usefulness for astronomical research would therefore have long since ended.

The Scientific Advisory Council

In February 1957 a distinguished chemist, Sir Eric Rideal, sent me a handwritten letter in which he explained why he was anxious for me to accept an invitation to become a member of the Scientific Advisory Council (more formally, the Advisory Council on Scientific Research and Technical Development), of which he was chairman. This council, like the Aeronautical Research Council, advised on defense matters. My initial instinct was to refuse, but after negotiations with the chairman of the Aeronautical Research Council, I resigned as vice-chairman of the Air Navigation Committee and accepted the invitation from Sir Eric. I learned that the task of the radar detection of ballistic missiles had been allocated by the Ministry of Supply to the Royal Radar Establishment (the wartime TRE) and that very little action had been taken. It had neither the equipment nor the experience that we had already accumulated at Jodrell. We possessed data that no one else in the world had obtained and that were of great interest to those concerned with defensive radars. Without any delay the ministry agreed to equip the telescope with one of the most powerful transmitters in the country so that we could carry out further investigations—particularly on the confusing effect of the echoes from meteors and the aurora borealis—and obtain more information about the optimum wavelength to be used for this long-range radar work. That was merely an astronomical bonus from one of the most extraordinary involvements of my career.

The responsible ministry (that of supply) seemed peculiarly lethargic about the dramatic development of Sputnik and the implications for the defense of the West. In February 1958 I wrote directly to the minister's scientific adviser (then Sir Owen Wansbrough-Jones) concerning a false alarm given by a radar in the United States. I sought to stimulate interest in this issue and produced a report for the advisory council on the missile detection problem.

For another year we continued to make the occasional radar investigations of the Sputniks and their carrier rockets and published a number of papers on the characteristics of the echoes. That might well have been the extent of our involvement had the defense correspondent of the *Daily Telegraph* not published a column in the paper of 19 January 1960 under the headline "Missile Early Warning Unit for Britain." He had obtained confirmation from the Ministry of Defense that an American ballistic missile early-warning system (BMEWS) was to be built in Britain. Instantly I sent this to the chairman of the Scientific Advisory Council, listing the various points we had constantly discussed in council and arguing that, until technical details were available, it was of the "very greatest importance that the SAC should not lend the weight of its scientific authority to the installation of a system which because of a likelihood of confusion might lead to errors of judgement which could have the very gravest military consequences."

I was dismayed by the official response that the question was being held in abeyance until "certain new advisory arrangements were set up." However, I was wrong in my assumption that nothing further would occur. A few weeks later I received a letter from Sir Solly Zuckerman, who wrote from the University of Birmingham, where he was the professor of anatomy. He was also at that time chairman of the Defense Research Policy Committee and chief scientific adviser to the secretary of state for defense. Zuckerman expressed concern and said, "What people like myself want to know is whether a defense system is or is not scientifically conceivable."

Zuckerman wanted authoritative advice and to that end convened a small panel. At one of these meetings in the autumn of

1960 I asked for details of the American equipment that was to be installed in Britain, for which the government was paying a large sum of money (the installation was then estimated to be costing £43 million). The appropriate government scientist replied that the scientific adviser to the Air Ministry did have some knowledge of the design but that it was insufficient for him to give a complete answer to my various questions. Neither had the government's independent Scientific Advisory Council been informed of the developments nor asked to comment on the suitability of the equipment.

Meanwhile the Air Ministry made occasional indecisive approaches to me about the use of the telescope in an emergency. Suddenly the matter achieved a degree of urgency. Everyone involved now realized that the telescope was still the only instrument in Britain (and indeed in Europe) capable of detecting a ballistic missile by radar. The American equipment now being installed at Fylingdales, in Yorkshire, was behind schedule. The civil engineering work had been bedeviled by strikes. Reports in the British newspapers alleged that the U.S. defense chiefs were bitterly disappointed over the delays and had wanted to use American labor on the site but that the British government feared that this would lead to serious trouble with the unions.

In October 1960 arrangements had been made for additional equipment to be installed at Jodrell and for a number of Royal Air Force mechanics to be trained in its use. But Jodrell was a part of the university, and I became anxious that serious objections might be made to this involvement. Fortunately the vice-chief of air staff readily agreed to write a letter to the vice-chancellor to clear the collaboration in general terms. Our code name was Project Verify, but Lord Hailsham, who had been appointed minister for science and technology, insisted that there be no "cover story" about our association.

Although initially surprised at this "no cover" decision, I was soon to be thankful. On 5 September the *Daily Express* appeared with the banner headline "Jodrell Is Ready for Khrushchev—Telescope Called Up to Beat Rockets," over an article by Chapman Pincher. That extractor of secrets had somehow obtained the details of my arrangements with the air staff:

> Agreement to use the famous radio telescope at Jodrell Bank, Cheshire, to detect enemy H-bomb rockets has been secured by the Government. . . . [The telescope] will stand in for the big early warning system on Fylingdales Moor, Yorkshire, the building of which is being seriously held up by strikes. It will be the nation's main warning system for at least a year. . . . This will give the RAF the four-minute warning it needs to get its bombers off the ground for a retaliatory attack. It will give the United States at least 15 minutes.

I never discovered how Pincher secured his information. The vice-chief of air staff, who had been my senior contact in the arrangements, seemed quite unconcerned. Indeed, it is possible that this was an inspired leak designed to calm public opinion in Britain and America at a time of increasing tension between the West and the Soviet Union. We remained in a position to be activated by the declaration of a "state of military vigilance" until late in 1963 when, at last, the BMEWS installation at Fylingdales relieved us of our burden. When the Cuban missile crisis of October 1962 was resolved I had perhaps more reason than many others to be thankful, but during those days particularly any idealism that the telescope had been built solely to study the remote regions of the universe vanished forever. In the decade during which my pleas to the government's scientific advisers for help in building the telescope had been rebuffed, the wheel had turned full circle.

Several years of most amicable contact about the defense use of the telescope ended in 1963. In August the commander in chief of Fighter Command took me by helicopter over the Yorkshire dales to Fylingdales. The equipment was approaching operational status, and I was glad to see these complex installations, which were shortly to take over the role of the Jodrell telescope, and the computerized analysis of the data that was to overcome the dangers of confusion about which I had complained many years earlier. Nearly a year later, in the summer of 1964, the senior staff of Fylingdales made a return visit to Jodrell, and that, coinciding with a pleasant letter of thanks from the minister of defense, finally ended my official contacts with the ministry and committees charged with the defense of the realm.

Cold War Consequences

These events in my association with the UK defense interests occurred during the years when the cold war between East and West was at its peak. Everywhere this confrontation produced a deep distrust of motives, and nowhere was this more acute than in scientific and technological activities. The pursuit of astronomical research is an essential international activity; in seeking to establish close collaborative arrangements with Soviet and American scientists, we had merely followed the natural instincts of our profession. In the event, in this cold war atmosphere the Jodrell telescope emerged as far more than a purely astronomical research instrument, and for many years I often felt sandwiched between the Americans and the Russians. In this era a team of Americans was working with us to use the telescope as part of the American space tracking network, and simultaneously we were involved with the Soviets and their deep-space network. There were occasions when the sandwich led to a political confrontation, such as in February 1964 when a military emissary from Washington arrived at Jodrell to forbid us to carry out an internationally agreed collaboration between the United States and the Soviet Union by means of the *Echo 2* balloon satellite, in which the telescope was the vital connecting link. With a flash of inspiration J. G. Davies, who was with me, said, "But the moon does not belong to the Americans; we will transmit to Gorki tonight via the moon." That is precisely what we did, and the opposition to the agreed tests via the *Echo 2* balloon satellite crumbled instantly.*

Despite these good relations between ourselves and the Soviet and American scientists, there were disturbing illustrations from both sides of the distrust occasioned by the cold war. There was, for example, the communication from a friend in America who sent me a copy of a long article in an American magazine that accused me of having Communist sympathies and associations. The writer had made a careful study of my life and had selected any conceivable incident that he could twist to support his thesis—such as my perfectly normal scientific contacts with Russian

*A full description of this incident can be found in chapter 15 of my book *Out of the Zenith*.

scientists. This came at a time when I was closely involved with the affairs of defense described in this chapter and sent me into a torment of anxiety. I asked the vice-chancellor if action for libel could be taken, but in his wisdom (he was a lawyer by profession) he asked me first to seek advice from my friends in America. At the time I was a member of the board of visitors to Harvard Observatory, and our elder son was studying geology at Harvard, so I consulted one of my oldest and most trusted American friends, Fred Whipple of the Smithsonian Astrophysical Observatory. His advice was firm. The article was in a magazine of the John Birch Society, and my taking legal action would give its members precisely the publicity they wanted. "If it is any comfort to you," he said, "an exactly similar article was written about Eisenhower when he was president."

On some occasions our legitimate activities have been castigated by both East and West. There was, for example, the historic occasion early in February 1966 when the Soviet *Luna 9* transmitted the first photographs from the surface of the moon. In the early afternoon of 4 February the facsimile transmissions from this probe received by the telescope were fed into a borrowed newspaper picture converter and produced, line by line, before our astonished eyes, a photograph of the rocky lunar surface with a part of the protective fairing of the probe in the foreground. The entrance hall and corridors of the building were jammed with newspapermen and photographers, and apart from force majeure there was no reasonable means by which I could keep this photograph from them. A hundred cameras flashed, and this photograph filled the front page of every newspaper in Britain the next morning.

The praise showered upon us for obtaining the first photograph ever published of this close-up view of the moon's surface was soon overwhelmed by a storm of abuse and criticism that descended upon us, but the suggestion that we were *intentionally* "scooping" the pictures was absurd. We were never to discover what bureaucratic procedures held up the release of these pictures in Moscow.

Of my visits to the Soviet Union, I write with reserve. There was initially a mutual admiration. The Soviets seemed as astonished that we could build a telescope as I was that they could launch a Sputnik. There was a suggestion that I go to the embassy in Moscow as the UK scientific attaché. I did not want to, and I am thankful that no pressure was exerted, but there was pressure of a different kind in 1963. In 1961 when the Russians lost contact with their first space probe to Venus, we responded to a request from the Soviet Academy of Sciences that they might be allowed to use the highly sensitive telescope to search for this lost probe. Shortly afterward I had a pressing invitation to make a return visit and did so in 1963. I was taken to their own new deep-space tracking installation—a massive affair on the Black Sea coast—the first foreign visitor. Afterward M. V. Keldysh, who was then the president of the academy, said that if I would stay with them they would build the huge telescope for which I was then seeking funds without success. I responded that I was an Englishman, and my proposal that the physicist Peter Kapitza* be allowed to visit us got a very dusty answer. Perhaps, as an American official suggested to me, they did not want me to return to England, and perhaps it was foolish of me to have gone to the Soviets when they must have known that we were then an emergency part of the Western defense system. Now, many years after these events, I realize how suspicious the Soviet hierarchy must have been of my motives, but my contacts were always professionally scientific and without hidden sinister motives. In my relations with the various departments of government in the United Kingdom and with my Soviet and American colleagues, I have always striven to behave as a good citizen. Nevertheless, the paradox of my relations with the Soviets was forcibly revealed to me some time later. One afternoon a member of the Soviet ambassador's staff in London telephoned to congratulate me on my election as a foreign academician of the Soviet Academy of Sciences. But this coveted honor never materialized, my inference

---

*P. L. Kapitza (1894–1984) made important contributions to low-temperature physics while working in the Cavendish Laboratory from 1921 to 1934. He was detained in Moscow during a visit there in 1934 and, during this time, the Institute for Physical Problems was created for him. In 1946 he was dismissed from his Institute. In 1954, after the death of Stalin, he was restored to favor and for several years had great influence in the Soviet hierarchy although he was not allowed to travel outside the Soviet Union until 1965.

being that my election was canceled in the final stage by a highly placed Soviet official.

In later years, as the atmosphere of the cold war softened, the tensions of my relations with the Soviets lessened. In the brush with Keldysh in 1963 they were at their peak. I had thought I would not see England again and only slowly recovered from the abnormal fatigue and depression that no medicine could cure. At the height of public approval it was as though all life had turned to dust and ashes. On my fiftieth birthday I talked of giving up science and entering the church. I did not do so, but those events had a lasting effect on my life and thought.

23

Diversions from the Workbench: The Myth of Power

The diversions to national problems of defense described in chapter 22 owed their origin to my work in World War II and to the potential of the telescope in a defensive role. However, simultaneously other diversionary activities began, of a type that affects almost all working scientists at some stage of their career—the allocation of the national budgets for research.

Until 1965 the UK government supported fundamental research in universities mainly through the Department of Scientific and Industrial Research (DSIR). I had served as either a member or the chairman of a DSIR subcommittee on astronomy since its establishment in 1958. Our meetings were short, informative, and enjoyable. We met only quarterly and could deal with our business in an afternoon. The application for money did not greatly exceed the sums available, and there was time to learn the details of other people's astronomical research projects. These pleasant administrative diversions were soon to become memories of a bygone age, before the judgment of the independent scientific advisory body was shaped by other political and bureaucratic interests.

327

During the summer of 1964 the Conservative government announced that there was to be a fundamental change in the arrangements for the support of research in Britain. Hitherto, in addition to the finance administered by the DSIR, various other organizations made substantial contributions to fundamental research. Now the whole of government money for civil research in physics, chemistry, astronomy, engineering, and space was to be administered by the new Science Research Council (SRC). When Melville, with whom I had recently served on the DSIR committees, summoned me to London on 22 December to discuss an urgent matter, I was apprehensive that he might seek to involve me in this new council. My fears were well founded. I could not share his opinion that he was honoring me by this request to help in the administration of the £25 million annual budget that would be available for the new council. It seemed to me that the job was nearly full-time, especially as Melville planned to establish me as the chairman of the board responsible for the grants for research in astronomy and space. I pleaded for time to think and consult the vice-chancellor. Sir William Mansfield Cooper, who had succeeded Stopford as vice-chancellor in 1956, saw nothing but good in Melville's proposals. He said it was a duty and could not understand my reluctance to accept this position of power, especially as I was already asking for substantial grants to construct another very large telescope at Jodrell. So in January 1965 I agreed to accept Melville's offer and in the spring of that year became a member of the council and the chairman of the Astronomy Space and Radio Board.

After two years as chairman of the Science Research Council, Melville decided to resign. It was not easy to find a successor in this burdensome office. The independent members of the council were asked if they would be willing to be nominated. When Melville turned to me, his question was not "Are you interested?" but "I don't suppose you're interested in the job, Lovell?" He knew that I already felt the truth of Wordsworth's lines "The world is too much with us; late and soon,/Getting and spending, we lay waste our powers." It was the autumn of 1970, five and half years after the initial meetings of this new body, before changed circumstances allowed me to retire honorably from this seemingly endless maze of meetings, arguments, and reports. My initial response to Melville that it was nearly a full-time job was a correct judgment. No man could spend half the week in the stuffy committee rooms of London and simultaneously carry out duties as a senior member of the university staff and direct the increasingly complex researches and developments at Jodrell.

The concept that any such position is one of power is a dangerous myth. It is the permanent civil service staff of such organizations that ultimately have some vestige of power, because the control of the budgets is their full-time responsibility. Their safeguard is, in the end, the independent members, who often have to acquiesce in decisions not necessarily based on their scientific judgment. Some of my academic colleagues soon recognized this danger. Jack Ratcliffe, from the Cavendish, felt so strongly about it after a few meetings of the board that he resigned on a matter of principle. Later Fred Hoyle, chairman of one of the committees responsible to the board, repeatedly urged me to resign, but I could see no good in this easy escape from responsible positions.

It is not my intention to attempt to write any detailed account of these years. Indeed, it would be the history of the development of much of British research in the physical and engineering sciences in that critical period when the government injected a new element by constantly pressing the council to spend its money mainly on research that was relevant to the immediate needs of the state. As far as my own developments at Jodrell were concerned, the vice-chancellor's belief that my presence as chairman of the appropriate board would facilitate the award of the substantial grants I was seeking proved to be a complete fallacy.

Now, more than twenty years later, I think of the wisdom of Francis Bacon, who nearly four centuries ago wrote, "It is a strange desire to seek power and to lose liberty."

The Academies

In chapter 22 and earlier in this chapter I wrote about diversions from the normality of academic life and research. Particularly in earlier days many academic scientists would refuse enticements to be so diverted, on the reasonable grounds that they would be seriously distracted from their research and teaching. During recent years the increasingly pragmatic attitude of the Western governments to scientific research has led to significant changes in this attitude, and pressures on academic scientists to engage in advisory and consultative work away from their own laboratories are now commonplace. Nevertheless, the choice remains with the individual, and, contrary to what the foregoing may imply, I have had occasion to reject offers of appointments that, although they may have been lucrative, I judged to be irrelevant to or incompatible with my appointed task.

An entirely different category of extramural activities, from which scientists will rarely turn aside, is the opportunity to serve the learned institutions concerned with their own disciplines. In the postwar years it has been my good fortune to be associated with several such academies, particularly the Royal Society and the Royal Astronomical Society. I imagine that, for any scientist of British citizenship, election as a fellow of the Royal Society is the most coveted event of a lifetime. My own election came in 1955, and there is no doubt that this event helped me, and enabled others to help me, survive the near disasters of the ensuing years. Since that time I have not, I think, rejected any request to help on committees of the society or in other activities, and these associations have been among the most pleasant of my scientific career.

My other main association is with the Royal Astronomical Society—the body that was of such critical importance in the early years at Jodrell and in organizing the scientific support for the telescope. To become a fellow of this society one merely has to be interested in astronomy and be proposed and seconded by existing fellows. The result is that the monthly meetings (always on the second Friday) are nearly always packed with enthusiastic young astronomers. My two-year term as president, 1969–71, embraced the 1970 sesquicentenary year of the society, providing the opportunity for several special functions and for the issue of a celebratory stamp by the British Post Office.

Internationally the interests of scientists are covered by particular unions with worldwide membership. Two have been especially relevant to Jodrell and my own work—the International Scientific Radio Union (URSI, Union Radio-Scientifique Internationale) and the International Astronomical Union (IAU). The general assemblies of these unions take place in various parts of the world every few years and are important meeting grounds for the appropriate international scientific communities.

These national and international diversions from the workbench are an essential part of the career of a professional scientist. Unlike the grant-giving bodies, they are almost wholly concerned with the progress of the science and—as I will next describe—with the protection of the disciplines. Indeed, my participation in these national and international organizations more than once led me into controversy over issues of importance to astronomy at large.

Environmental Strife

Radio astronomers are probably in more danger from the destruction of their terrestrial and space environment than any other group of scientists. Since I arrived at Jodrell Bank, and especially since the building of the telescope began, I have had to spend many weary hours in combat with persons whose interests are antagonistic to those of our researches. The radio signals that reach us from the distant regions of the universe are exceedingly weak by terrestrial standards. It has been estimated, for example, that the total energy collected by radio telescopes since the science began would be barely sufficient to light a flashlight bulb for a second. It is therefore easy to appreciate that terrestrial signals are a serious form of interference to those who use radio telescopes. These continuing terrestrial troubles were punctuated in the early 1960s by two events in space that forced us into the public gaze.

The Needles or West Ford Project

Early in September 1960 the Thirteenth General Assembly of the International Scientific Radio Union (URSI) convened in London.

At that time I was chairman of the commission on radio astronomy, and I vividly recollect the consternation caused by an American scientist, W. E. Morrow, Jr., when he described a proposal by MIT's Lincoln Laboratory to place in orbit a huge number of thin wires in order to test a new method for secret communications.

In essence the idea was to encircle the earth at a height of a few thousand kilometers with a reflecting belt of copper wires. The radio beam from a powerful transmitter in one part of the world would then be directed at this wire belt, and, provided all the parameters were known, the signals would be picked up by an equivalent receiving system in another part of the world. Before the days of synchronous-orbit communication satellites, this was specifically designed for instant worldwide communication by the U.S. defense authorities. The initial plan was for a test launch of 35 kilograms of very thin dipoles 1.7 centimeters in length. This package would contain 350 million of the wire dipoles; once the payload reached orbit, these dipoles would be released from a rotating dispenser so that a cloud of them would encircle the earth.

Such belts of dipoles would scatter sunlight into the dark areas of the earth, interfering with optical astronomers and also scatter radio transmissions from distant sites into radio telescopes. The prospect that a considerable number of such belts would be maintained permanently around the earth instantly led to violent protests by the astronomical community. The commission of which I was chairman immediately drafted a resolution calling upon URSI in conjunction with the IAU to take urgent steps to "ensure that such projects cannot endanger future astronomical research."

When in August 1961 it was announced that the U.S. Air Force had received presidential approval to proceed with the project, a further anxiety had developed. If the dipoles were released in the planned orbit, they would exist in space for only a few years. Unfortunately, no safeguards had been introduced to ensure that release would be made only if the planned orbit was achieved. Release in a different orbit might lead to lifetimes of hundreds of years, with unforeseeable consequences for the future of astronomical observations.

The nature of the exchanges left little doubt that the U.S. administrators were sensitive about the proposal. In response to frequent inquiries from the press I had expressed concern that, since this would be a poor form of communication compared with the communications satellites then under development, there must be more sinister defense motives in carrying out these tests. Early in September 1961 the president's science adviser, Jerome Wiesner, wrote to me from the White House in an attempt at reassurance, but neither his letter nor his subsequent telephone call quelled the anxiety that the United States was about to set a precedent in the contamination of space that would have grievous consequences. In the United Kingdom scientists continued to urge the United States not to release the needles until it was certain that the correct, short-lifetime, orbit had been achieved. It was too late, but ironically fate intervened. At about two in the afternoon on 21 October 1961 the West Ford container was launched from Point Arguello, California, without any safety device. The container was successfully injected into orbit. The 350 million wires were embedded in a cylindrical block of naphthalene about 45 by 15 centimeters in size. The wires should have been released from the naphthalene binder as it warmed up, but they were not. Subsequently it was concluded that the naphthalene block had broken up into several pieces but that no wires had been dispensed into space.

The long delay occasioned by the need to investigate the failure of this launch gave time for wiser counsels to prevail, and when a successful payload was eventually launched, on 9 May 1963, a safety device was installed so that the needles would not be released until it was established that they would dispense in the correct orbit. In the event, the orbit was correct and the command to dispense the package was given on 10 May. Two days later the dipoles were first observed in orbit by the Lincoln Laboratory radar.

A year later full accounts published by the American scientists indicated that about one-quarter to one-half of the 480 million dipoles in the package were orbiting as individual dipoles. The remainder were orbiting as undispensed packages of various sizes. Fortunately the advance of space techniques in the form of synchronous-orbit satellites soon gave the military authorities superior means of instant worldwide communications, and the prospect of the multiplicity of long-enduring belts of wire in orbit around the earth that caused such dismay among the world's astronomers has never materialized.

The Rainbow Bomb

In 1962, in the midst of the arguments about the needles, I became involved in an acrimonious dispute concerning the intention of the U.S. Defense Department to explode a nuclear bomb at an altitude of several hundred miles in a most sensitive region of the earth's ionosphere and radiation environment. In March 1959 the *New York Times* had reported that in the preceding summer the Americans had exploded a nuclear device in space. After this publicity the U.S. defense officials confirmed the story of Project Argus.* Bombs with a yield of one or two kilotons had been exploded three hundred miles over the South Atlantic. The official release stated that the "military results were somewhat negative" and that "in a sense the results were mostly scientific." In fact, the "scientific results" were alarming. The earth's ionosphere was temporarily disrupted, and there were displays of artificially produced aurorae, magnetic storms, and a fade-out of shortwave radio communication. The electrons produced in the explosion moved out into space and were trapped in the earth's magnetosphere, producing an artificial radiation belt comparable in intensity to the natural radiation belts discovered by James Van Allen in the first U.S. Explorer satellite early in 1958. After three months this radiation belt was still detectable.

The announcement in April 1962 that the Americans intended to carry out further tests of the effects of explosions in space with a megaton nuclear device, a thousand times more powerful than the Argus explosives, caused the utmost dismay. There were two issues at stake. Both Russian and American official scientific

*The first high-altitude explosion was made in August 1958. The Argus series followed in August and September.

administrations had recently subscribed to resolutions of the international scientific unions declaring that no group had a right to change the earth's environment in any significant way without full international study and agreement. The second issue was simply the belief that a megaton explosion at an altitude of several hundred miles would indeed change the earth's environment—in particular, that the effects on the earth's radiation zones would be serious.

The strong views of many British scientists about these intentions were paralleled by the reactions of many in America. With the press on my heels, I wrote in the London *Observer* of 6 May that, if the Americans had data that convinced them that the damaging effects would be temporary, they should "produce the information before they make this sledge-hammer blow at the environment of the Earth." Questions were raised in the House of Commons, and in mid-May I was asked to prepare a memorandum on the subject for Prime Minister Harold Macmillan. Before the end of the month he wrote to say that he had taken steps to see that the attention of the appropriate U.S. authorities would be drawn to the dangers.

Neither these representations from the United Kingdom nor the growing alarm of sections of the American scientific community and public had any effect, and at 09^{h}00 Universal time on 9 July 1962 a megaton bomb was exploded at a height of 400 kilometers over Johnston Island, in the Pacific Ocean. The effect was cataclysmic and had consequences even beyond the pessimistic estimates. The upper ionosphere layer around the earth was broken up, causing a severe disruption of long-distance radio communication. This was accompanied by auroral displays and magnetic-field disturbances. The electromagnetic pulse (EMP) from the explosion sent massive electrical currents through the power lines in the Honolulu area, some eight hundred miles distant. Strings of street lamps were burned out, and circuit breakers were opened over a wide area. These temporary effects were insignificant compared with the long-term changes in the radiation zones. In an official report issued by the Atomic Energy Commission, the Department of Defense, and the National Aeronautics and Space Administration on 20 August, it was estimated that more than a hundred trillion trillion electrons from the fission product decays had moved out into space and been trapped in the earth's magnetic field to form a new radiation zone at an altitude of about 4,000 kilometers with an intensity more than a hundred times greater than that normally observed. From the rate of decay that had occurred in the few weeks following the explosion, it was estimated that the main artificially created belt of trapped particles could have a half-life of twenty years. The solar cell power supplies of several satellites had been seriously damaged, and on 12 July the British satellite Ariel ceased operating. Measurements made in Peru for several months after the explosion showed that the rate of decay was slow and that about 10 percent of the excess radio noise would still be present after two years.

The arrogant claims of the American scientists that the effect of the megaton Starfish explosion at high altitude could be extrapolated from the low-level kiloton explosions in the Argus series were completely unfounded. We did not claim that we could predict the effect of the Starfish explosion, but we did claim that the Americans could not predict it either. At least those of our colleagues who had poured scorn on our worries had the courtesy to apologize.

The political effects were considerable. The intense radio emission from the trapped electrons was of particular concern to Ryle and myself, and in October I communicated once more with Wiesner at the White House to complain that if this series of explosions continues in space, and at different latitudes, "the outlook for a whole series of radio astronomical measurements from Earth becomes extremely bleak." In his reply he wrote, "While it must be recognized that the decision to conduct high-altitude tests was based on military requirements, I can assure you that careful consideration has been given to the yields and altitudes of the remaining high-altitude tests in our current series." In fact, in a press conference on 13 September, President Kennedy had already given an assurance that the altitude and yield of any remaining tests in the series would be reduced.

The controversy continued, and in May 1963 the UK government published an official document conveying our recommendations that no country should take any action that would cause a major disturbance of the natural environment of the earth without prior consultation with other countries. In the event, no further high-altitude explosions occurred. In our private correspondence with the prime minister and the minister for science, and in the exchanges in the House of Commons, the emphasis was always that the only safeguard lay in the proposed test ban treaty. Some commentators subsequently maintained that the publicity given to the Starfish explosion by the British attitude had a significant bearing on the agreement and signature of the nuclear test ban treaty in the Kremlin on 5 August 1963. If that is indeed true, it would be a happy recompense for an unpleasant interlude in our careers.

Terrestrial Interference

Although the West Ford and Starfish incidents were dramatic and represented immediate dangers for radio astronomy, terrestrial interference poses a very serious cumulative threat. This interference takes two forms. One arises from the radio transmissions used for broadcasting, television, private and commercial radio telephones, and endless other purposes. When we commenced our work after the war, large regions of the spectrum were free of such transmissions, and we had a wide choice of wavelengths to which we could tune our receivers and study the weak emissions from the universe without interference. This environment deteriorated with alarming rapidity, and quite soon we had to seek national and international agreements with the appropriate regulatory authorities to grant us a few protected wavebands on which no other user of radio communications was allowed to transmit. The international unions (particularly URSI and the IAU) formed special and joint committees through which agreement was sought for the protection of a few vital bands in the spectrum for our use. This problem is worldwide, and as the use of radio becomes more and more widespread, the radio astronomer's environment is being destroyed.

The second form of terrestrial interference is of an impulsive nature, arising from electric sparks and discharges—the unsuppressed ignition from car engines, for example, is a notorious source. The curvature of the earth means that this is a relatively local nuisance, and radio astronomers seek uninhabited areas of country in which to locate their instruments. Initially Jodrell was isolated, but the growth of traffic in the years after the war and the spread of new housing developments caused serious trouble. Like our counterparts at many other such establishments, we had to seek agreements with the planning authorities to minimize the spread of these developments and the associated traffic within a several-mile radius of Jodrell. In *The Story of Jodrell Bank* I have described some of the troubles we encountered and the unfortunate publicity that often arose from the public inquiries.

In these matters the voice of pure research is very weak indeed when in combat with commercial and military interests. In fact, in moments of despair I have said that the pursuit of radio astronomy on earth may well prove to be a phenomenon limited to the second half of the twentieth century.

24

My Life and Thought

At the beginning of one Easter vacation when I was an under-graduate student in Bristol, I wandered rather aimlessly into the main library, seeking some divertissement to fill the empty days until the term began again. Fate guided me to Albert Schweitzer's *My Life and Thought*. The English translation did not appear until 1933, and I suspect that I saw it because it was on the new-acquisitions shelf. In any event it was a fortunate encounter. The prospective dullness of the Easter vacation was suddenly transformed by the shining example of Schweitzer's mastery of a range of intellectual disciplines.

A philosopher, a professor of theology who transferred his career to medicine in order to become a missionary, a pupil of Charles Marie Widor who saved the old organs in Europe from modernization, who wrote two authoritative volumes on Bach, and who in his recitals performed Bach with a new insight. For me in my twentieth year these revelations of human accomplishment were an inspiration. I studied his Bach (and still do so more than half a century later). I returned to the library and became absorbed in his 1922 Dale Lectures in Oxford, published as *Civilization and Ethics* and *The Decay and the Restoration of Civilization*. I did not know what to make of his theological writings, and so I gave my father *The Mysticism of Paul the Apostle*. His fierce reaction

339

first made me aware that theology and religion can be vastly different subjects.

It was the electric arcs produced by Tyndall that turned me into a scientist, and it was the reading of Schweitzer's *My Life and Thought* a few years later that kept me from a complete and absolute absorption in science. I became Tyndall's pupil but never met Schweitzer. And that is a lasting regret, for I could easily have done so. In 1960 Schweitzer was eighty-five years old and about that time made a brief visit to London. He stayed with my wife's uncle, who, as a medical missionary in the Belgian Congo between the wars, had known Schweitzer well. Now, as Sir Clement Chesterman, he worked in London as an authority on tropical diseases and the president of the Schweitzer Hospital Fund. On an autumnal Saturday evening after I had lectured in London, to my everlasting regret I accepted the offer of a lift home instead of the hospitality that would have brought me close to Schweitzer. Soon afterward Schweitzer returned to Lambarene, and all that I could do thereafter was to play a little Bach on the piano that now bore the inscription "Dr. Albert Schweitzer played on this piano."

A few years ago, more than fifty years after my awakening in that Eastertime, I was asked to address an International Albert Schweitzer Symposium on the theme "Reverence for Life and the Cosmos." I realized that in the intervening years my thought had departed radically from the views expressed by Schweitzer. I could no longer agree with his statement "Today thought gets no help from science—the newest scientific knowledge may be allied with an entirely unreflecting view of the Universe—the co-ordination of the different branches of knowledge and the utilization of the results to form a theory of the Universe are, it says, not its business."* But I also realized that the query he posed in *The Decay and the Restoration of Civilization* "Is it possible to find a real and permanent formulation in thought for a theory of the Universe which shall be both ethical and affirmative of the world and of life?"† had remained at the core of my own fascination with

---

*Albert Schweitzer, *The Decay and the Restoration of Civilization*, trans. C. T. Campion (London: A. & C. Black, 1932), p. 72.
†Schweitzer, *Decay and Restoration*, p. xi.

the interfacing regions of theology, philosophy, and science. Throughout my career as a scientist I have frequently spoken and written about my evolving views on these transdisciplinary issues and have regarded this as an important part of my work. In view of the life I have described in this book, this is perhaps surprising, but science has never claimed the whole of my existence.

In 1937 and 1938, for example, as the cataclysmic events were unfolding in Europe, part of my mind was wrenched away from the interpretation of the cosmic rays in my cloud chamber to think about the real world. I wrote *Science and Civilization*. It belongs to the 1938–39 world in which it was published, revealing the wrath and bewilderment of a twenty-five-year-old at the frustration of science—at the reality that science was so often diverted to destructive purposes instead of to human welfare. As I look at those pages now, I see with profound misgiving that the underlying thesis remains valid. With starvation and illiteracy rampant in the world, the frustration remains. The lopsided and pragmatic use and intepretation of the scientific discipline have not changed. The names, the figures, and the topics have changed but not the corporate antagonistic attitude to the physical and intellectual survival of the human species.

One reviewer wrote of the "high moral tone" of that book. It was a tone soon obscured in the caldron of the war—but not completely. Recently a historian, searching my wartime records, discovered a copy of an address I had given in TRE in October 1944. It seems that A. P. Rowe, the head of the establishment, had posed the question how, in peacetime, we could attain for the betterment of mankind the same spirit of self-sacrifice and of living for a cause that we had in the war. I had forgotten this address; neither can I recall the circumstances that led to this wartime assembly at which I tried to answer Rowe's question. I sought the causes. First came the belief that our age is living under the sign of the collapse of civilization. The second dealt with the contemporary effects of this collapse: war, hunger, disease, frustration, and I expressed the conviction that the really *fundamental* cause of the progressive decay of civilization is the inadequacy of its ethical basis.

The Reith Lectures

For twelve years after the war my thoughts were dominated by the establishment of Jodrell Bank and the building of the telescope. Then, when the difficulties were most intense, I received a letter that sustained my hope for the future and that was to mark a turning point in my outlook on the place of science in human thought. On 25 February 1958 General Sir Ian Jacob, the director general of the British Broadcasting Corporation, wrote to invite me "to give the BBC's Reith Lectures in the late autumn of the year." He explained, "Our aim in broadcasting them is to promote an opportunity for someone in the van of contemporary thought to make an individual contribution to a subject that should receive wide and serious attention."

It was I think the most unexpected and surprising invitation I have ever received. The Reith Lectures were named after Lord Reith, the first director general of the BBC, and since their inception in 1948 had been given by a very distinguished sequence of lecturers. The first lectures were delivered in 1948 by Bertrand Russell, on the topic "Authority and the Individual." J. R. Oppenheimer delivered the lectures in 1953, entitled "Science and the Common Understanding," in 1956 Appleton spoke under the title "Science and the Nation," and the 1957 lectures, "Russia, the Atom and the West," were delivered by George F. Kennan. In those days television had not yet gripped the national interest, and the broadcast of the spoken word attracted a large audience, particularly at a peak listening time after nine on Sunday evenings in the late autumn when the lectures were to be delivered. The letter from Sir Ian Jacob that arrived on my desk at Jodrell on that February morning suddenly appeared as a demonstrative reaffirmation of faith in me as an individual and in the project that was tearing my life asunder.

In his letter of invitation to deliver the lectures Sir Ian Jacob expressed the hope that I would be attracted to develop the themes of some of my recent broadcasts. I had over the years given a number of broadcast talks on the telescope and on our present knowledge of the universe, and the development of those themes presented no particular problem. However, I knew that

342

Sir Ian also had in mind the more fundamental problems concerning man's relation to the universe that had been the theme of a talk broadcast on Christmas Eve four years earlier, before the major troubles with the telescope limited my horizon.

That talk, "Man in the Universe," marked the first occasion after the experiences of the war on which I had given any expression to the conflicts that had surrounded my life. The idealism of my youth, the intensity of my early religious surroundings, the iconoclastic impact of the war, and many of the implications of contemporary science were disturbing elements in my life. For that Christmas Eve broadcast I had to begin the process of resolution. "We have grown to fear science and to despair of religion. We fear science because we feel it might destroy us; and we despair of religion because we feel that it no longer seems able to prepare man to face the contemporary world."

I discussed the reasons for this tension and asserted that it arose because of the failure of our moral and religious teachers to adjust human thought to the ever increasing knowledge and power of science—and that this cause was always changing. In this century it concerned the problems posed by evolutionary theory, then those arising from the philosophical implications of modern science, and now the deeply rooted controversy about the origin of the universe. This last issue formed the main theme of the talk: the question of the uniqueness of man and the possibility of a scientific explanation of the origin of the universe.

Nearly five years later that theme was prominent in my Reith Lectures. The lectures took a tortuous path during their evolution. I know that some authors can think straight onto a tape recorder, or a typewriter, but I never have been one of those. The Reith Lectures were composed with pen and ink. Fortunately, the seven bound volumes of all the correspondence, the press comments, the typed drafts, and the final versions contain those original scripts. I have them before me, and as I turn over those manuscript pages I am reminded of the original scores of Beethoven's last sonatas—almost indecipherable because of the deletions, the insertions, and the rewriting. Even the drafts as eventually typed by the BBC bear witness to the continuing travail.

By May of 1958 I had agreed with Kenneth Brown, who had

been assigned by the BBC to be the producer of the lectures, that I would let him have the drafts of the first two lectures by the end of August and those of the remainder by the end of September. However, the drafts first had to emerge from the critics in my own home—always difficult but never more so than with these lectures. One Sunday evening when I was feeling quite pleased with my composition of the first lecture, Colonel Walter Hingston, the chief information officer of the DSIR, was staying at The Quinta. The warm fire induced a mellow atmosphere, and it was suggested that I read out the script of my proposed first lecture. The lectures were each of thirty minutes' duration, and throughout Walter Hingston and my wife listened with rapt interest—or so I thought. I finished the reading and waited expectantly. "Could I just see the manuscript?" asked my wife. When I handed it to her, she rose and cast it into the fire. "Now you can start again." Of course, she was right, and Hingston politely agreed that I would have lost half the audience within five minutes. The contents of that burned manuscript of one lecture contained all the salient points that were eventually to be spread out over the whole series of six.

The drafts subsequently reached Kenneth Brown, the producer, and the next sequence of incisive criticism began. When I returned from meetings in Moscow at the end of August, his response was waiting for me—over five single-spaced quarto pages dealing with my misconception or unawareness of commonly accepted theological views. He could not accept my recurring argument "that cosmological theories could, under certain circumstances, pose theology with a dilemma." Throughout that long letter he argued the theistic position that at no point could theology and science conflict, because the methods of inquiry and the presumptions were entirely different.

My response was instant and unequivocal:

> I am aware of the position you set out, but I find it completely unacceptable. In short I have no patience with those theologians who plug their ears and blindfold themselves in the face of science. I have made an attempt at the reconciliation and integration of the theological and scientific aspects of this problem. My effort will be pointless if I am to accept the particular doctrine that at no point do theology and science

conflict or interact. The idea that scientific cosmology can have no influence on theology I find quite untenable.

The conflict between Kenneth Brown and myself was absolute, and I had no further dealings with him. After a few days I received messages from the BBC that Mr. Brown was ill and that John Weltman would now be my contact and producer. Weltman argued fiercely enough with much of the content of my drafts, and throughout September and October there were many changes and a good deal of reorganization of the lectures, but there was no longer an irreconcilable conflict over the fundamental theological issues.

By good fortune the summer months of 1958 contained meetings in Brussels, Paris, and Moscow that provided me with stimulating experiences that were to exert a considerable influence on the lectures. This was a period of intellectual ferment in astronomy, particularly in cosmology, and of tension in international affairs. The lectures inevitably reflected that atmosphere. The title "Astronomy and the State" and much of the content of my fourth lecture emerged from my discussions with many of the American scientists who were in Paris and from my first and most memorable visit to the Soviet Union, which followed immediately. I described the development of the telescopes in Great Britain, in America, and in the Soviet Union and discussed the present possibilities arising from the recent launching of earth satellites. "A survey of the international scene in astronomy gives me excitement and pleasure as an astronomer," I went on, "but as a citizen, I am filled with dismay." This new era of astronomy involving observations from space vehicles would be impossible "but for the political and military divisions of the world which have forced the governments to an expenditure which would never be borne as a budget for fundamental scientific work alone."* In concluding that lecture, I referred to the pessimistic outlook that the scientific and technological superiority of the USSR over the West would soon be complete, but I added, "I do not believe that this will necessarily be the case, because I think

*A. C. B. Lovell, *The Individual and the Universe* (London: Oxford University Press, 1959), pp. 60, 64.

345

the restraints on freedom . . . may reduce the effectiveness of Russian science and may counteract to some extent their enormous superiority in scientific man-power and finance."*

Thirty years later, although some of the content of that lecture might have to be modified, the pessimism and the doubts remain. "The pursuit of the good and evil are now linked in astronomy as in almost all science," and, alas, it remains true that "the fate of human civilization will depend on whether the rockets of the future carry the astronomer's telescope or a hydrogen bomb."†

At the time of my visit to Moscow in 1958 rather little was known in the West about the postwar developments in Soviet science. It was therefore of special interest to me to learn about the astronomical work of the Soviet scientists. Again I was led to a substantial modification of my plans for the second Reith Lecture, on the origin of the Solar System. In Moscow it soon appeared that the Soviet theorists had in recent years developed a great interest in the problem. I discovered to my astonishment that the study of the origin of the Solar System by Russian astronomers was justified because of its importance in the elaboration of a correct materialist world outlook. "Cosmogony in the USSR," I was told, "is based on the firm materialist traditions of Russian science." I was near Fred Hoyle in the library of the Sternberg Astronomical Institute. He suggested that I search for a book on cosmology and assured me I would not find one. He was correct. A ban existed on the translation of cosmological works because their political philosophy restrained Russian astronomers from working on the subject. The failure to find a scientific solution to the cosmological problem might, it was feared, lead the Soviet investigators to the idea of God.

The Reith Lectures penetrated to the Soviet Union, and three years later Alla Massevitch, a prominent Soviet astronomer who was then in charge of the space tracking network, came to Jodrell to use the telescope in a search for signals from the lost Soviet probe to the planet Venus. I was on the platform at Crewe Station waiting for her arrival. She appeared, thrusting a book at me. "There, you silly man, how could you say such things? Read about our Soviet

*Lovell, *Individual and Universe*, p. 72.
†Lovell, *Individual and Universe*, pp. 72, 73.

conference on cosmology." The imposition of the political doctrine had been steadily evaded because increasing numbers of Soviet astronomers were visiting the West and returning with books on cosmology. By the 1970s an important theoretical school had emerged. The academician Ya. B. Zeldovich emerged as a distinguished cosmologist. He delivered the inaugural address at a cosmological symposium in Poland in 1973 and posed the question "What was the state of the universe in the remote past?" In answer he said that "the scientific investigation of these questions was based to an equal extent on logic and intuition, on hypothesis and rigorous proof" and that it depended "on individual prejudices, the likes and dislikes of authors, and perhaps even on their sub-conscious Freudian attitude to such things as order, chaos, anti-matter."* Neither this question nor the answer could possibly have been given by a Soviet scientist in 1958.

The Final Lectures—The Origin of the Universe

The last two Reith Lectures, on the origin of the universe, were to be given on the Sunday evenings of December 7 and 14, and in no other part of the original manuscripts do the processes through which the lectures emerged show more clearly. The problem was not merely to give a clear account of the cosmological problem as it was then seen. The real difficulties revolved around the rationalization of my own attitude and especially my increasing feeling throughout those months that there was not, and could never be, a final scientific solution to the cosmological problem. It was the experience of the Solvay Conference in Brussels, which that year was devoted to the theme "The Structure and Evolution of the Universe," that gave me the confident background to plan my lectures on the origin of the universe. At the time of my Reith Lectures the decisive observational evidence against the theory of continuous creation did not exist, and in those two lectures I contrasted this theory with the concept of the

*Ya. B. Zeldovich, *Confrontation of Cosmological Theories with Observational Data*, IAU Symposium, no. 63, M. Longair, ed. (Dordrecht, Netherlands: Reidel, 1974), pp. ix–x.

evolutionary universe expanding from a highly condensed state several billions of years in the past. It was to be another seven years before the discovery of the microwave background radiation, uniformly distributed across the sky, provided decisive evidence in favor of an evolutionary universe. The steady state, or continuous creation, theory, with the implication of a universe that had existed for an infinite past time and would continue to exist forever, was effectively abandoned. To the extent that observational cosmology has made this important advance since 1958 the description of the alternative universes that I presented then is no longer relevant. However, what I thought about the fundamental issues then as an ordinary human being has not significantly changed.

About our penetration into the history of the universe I said, "Finally, we shall reach a stage where theories based on our present conceptions of physical laws have nothing further to say. At this point we pass from physics to metaphysics, from astronomy to theology, where the corporate views of science merge into the beliefs of the individual." In our attempt to describe the initial moments of the universe, I continued, "we reach the great barrier of thought because we begin to struggle with concepts of time and space before they existed in terms of our everyday experience. I feel as though I've suddenly driven into a great fog barrier where the familiar world has disappeared. . . . Philosophically the eventual problem in the conception of the beginning of the Universe is the transfer from the state of indeterminacy to the condition of determinacy after the beginning of space and time when the macroscopic laws of physics apply." After enlarging on this point, I concluded, "philosophically, space and time had a natural beginning when the condition of multiplicity occurred, but the beginning itself is quite inaccessible."*

Although the observational picture has changed much in the intervening years, I would today repeat the closing words of those lectures:

If I were pressed on this problem of creation I would say that any cosmology must eventually move over into metaphysics for reasons which are inherent in modern scientific theory. The epoch of this trans-

*Lovell, *Individual and Universe*, pp. 93–111.

fer may be now and at all future time, or it may have been twenty thousand million years ago. In respect of creation the most that we can hope from our future scientific observations is a precise determination of this epoch. I must emphasize that this is a personal view. The attitudes of my professional colleagues to this problem would be varied. Some would no doubt approve of this or a similar line of metaphysical thought. Others would not be willing to face even this fundamental limit to scientific knowledge, although, as I have said, an analogous limitation occurs in modern scientific theory which describes the well-known processes of atomic behaviour. Some, I am afraid, will be aghast at my temerity in discussing the issues at all. As far as this group is concerned, all that I can say is that I sometimes envy their ability to evade by neglect such a problem which can tear the individual's mind asunder.

On the question of the validity of combining a metaphysical and physical process as a description of creation, this, as I said earlier, is the individual's problem. In my own case, I have lived my days as a scientist, but science has never claimed the whole of my existence. Some, at least, of the influence of my upbringing and environment has survived the conflict, so that I find no difficulty in accepting this conclusion. I am certainly not competent to discuss this problem of knowledge outside that acquired by my scientific tools, and my outlook is essentially a simple one. Simple in the sense that I am no more surprised or distressed at the limitation of science when faced with this great problem of creation than I am at the limitation of the spectroscope in describing the radiance of a sunset or at the theory of counterpoint in describing the beauty of a fugue.*

The Final Confrontation—The Limits of Science

If I had to deliver the Reith Lectures now, a great deal of the content would be changed, but my attitude to the ultimate confrontation would be the same. In the thirty years since I gave them, occasional events have continued to urge me to think and write about those misty areas of the mind where science tries to enter the domains reserved for theological dogma.

In 1975 I gave the presidential address to the British Associa-

*Lovell, *Individual and Universe*, pp. 109–10.

tion for the Advancement of Science. The title of my address, "In the Centre of Immensities," was taken from Thomas Carlyle's *Sartor Resartus:* "What is Man? who sees and fashions for himself a Universe, with starry spaces and long thousands of years . . . as it were, swathed-in and inextricably over-shrouded; yet it is sky-woven and worthy of a God. Stands he not thereby in the centre of Immensities, in the conflux of Eternities?"

The cosmology of this address was updated from the Reith Lectures. On the attitude of modern society I remarked, "The simple belief in automatic material progress by means of scientific discovery and application is a tragic myth of our age. Science is a powerful and vital human activity—but this confusion of thought and motive is bewildering to man, and it is a most alarming thought that the present antagonisms of society to scientific activity may deepen further."*

The science correspondent of *The Times* of London complained that my address was a return to moral philosophy. Maybe that was the case, but today there is nothing I would wish to withdraw from it. The antagonisms of society to scientific activity have indeed deepened further, to an alarming extent. I believe, too, with even more confidence, that the quest for scientific understanding is very far from embracing the totality of human purpose.

The decision of the *New York Times* to reprint this address was to have an important influence on the development of my thought. Shortly afterward I received a letter from Ruth Nanda Anshen in New York asking me to contribute to a series of books of which she was the founder and editor—World Perspectives. The challenge to develop the fundamental ideas of the lecture was one I could not resist, and the book with the same title as the lecture—*In the Center of Immensities*—was published in America in 1978. For me the gestation of this book was, like that of the Reith Lectures, a profound and disturbing experience. In my sixties I had the searing intellectual experience of discovering that I did not really understand the deeper meaning of much of the physics of

*Bernard Lovell, "In the Centre of Immensities," *The Advancement of Science,* 29 August 1975, p. 2. For a modified version of this address see "Whence," *New York Times Magazine,* 16 November 1975, p. 27.

the natural world that I had been teaching throughout my life. The attempt to understand the physical significance of the beginning of time and space almost became a metaphysical experience.

I started work on *Emerging Cosmology*, which, published in 1981, surveys the development of cosmological ideas over two millennia. Each age has believed that its contemporary view of the universe corresponds to reality. Nowhere is this more evident than during our own time, when striking discoveries in astronomy appear to provide scientific data about the condition of the universe in the earliest moments of its existence. By the observation of the microwave background radiation, we appear to have a scientific measurement that reveals the state of the universe some 99 percent of the elapsed time to the initial moment of its origin—commonly called the hot big bang. During recent years the theoretical and experimental researches in nuclear physics have had significant success in leading to at least a partial understanding of the forces and of the exotic particles that dominated the dense, high-temperature, radiation phase of the first million years.

These observations of the universe reveal a part of the natural world to which quantum theory and relativity apply, and it is a known world. The transference to the unknown world, where physical theories and scientific reasoning fail, inevitably leads to the fundamental query about the possibility of attaining knowledge. The vital question is whether, after the transference from the known to the unknown, what lies beyond contemporary scientific knowledge must forever remain a matter of belief or disbelief. In the known world we achieve scientific knowledge because we can proceed on the assumption that the object of the investigation exists independently of us. In these macroscopic cases we can externalize the object of investigation, be it a speck of dust, the moon, or a distant galaxy. When we seek knowledge about the initial state of the universe near the beginning of time and space, it is not evident that such externalization is possible. Here we have a special circumstance where the object of our investigation, although localized by scientific reasoning, is itself the entirety of time and space.

In an earlier cosmological context when the observations with the large telescopes first revealed the vast extent of the evolving universe, Alfred North Whitehead in his 1925 Lowell Lectures clearly foresaw this problem: "There is no parting from your own shadow. To experience this faith is to know that in being ourselves we are more than ourselves. . . . While the harmony of logic lies upon the Universe as an iron necessity, the aesthetic harmony stands before it as a living ideal moulding the general flux in its broken progress towards finer, subtler issues."*

The Nature of Belief

The writings of Karl Barth after the First World War have had an important influence on Protestant thought in this century. His insistence that God is transcendent, known to man only when He chooses to reveal himself (as in Jesus Christ) has had a vital influence on the attitude of many people in the sense that it is impossible for theology and science to interact. It is, perhaps, the easiest rationalization for the scientist, in which the starting point is not man's search for God but God's self-disclosure to man.

It seems to me that no purpose relevant to human existence will be served by denying that any contact exists between the two great forces that have shaped the modern world. Indeed, Whitehead himself is responsible for the major attempt of the twentieth century to include science and religion in a unified view of reality. In *Process and Reality* he developed the metaphysical system bringing aesthetic, moral, and religious concepts into relation with natural science. In the theology of Whitehead's process philosophy, God is a timeless source of order: "Apart from the intervention of God there could be nothing new in the world, and no order in the world."† In sharp distinction to neo-orthodoxy, Whitehead maintains that God both receives from the world and contributes to it. God is not transcendent, but He is immanent, omnipresent, and a universal influence. On the subject of creation Whitehead evades the problem of an ex nihilo event: "He is not *before* all creation but *with* all creation."‡

*Alfred North Whitehead, *Science and the Modern World* (London: Cambridge University Press, 1933), pp. 23–24.

†Alfred North Whitehead, *Process and Reality* (London: Cambridge University Press, 1929), p. 349.

‡Whitehead, *Process and Reality*, p. 486.

The liberal theologies of the twentieth century similarly stress the immanence and not the transcendence of God and accept that there is no fundamental separation of science and religion. God is not revealed exclusively by Jesus Christ but through many channels—created order, our moral conscience, and religions other than Christianity. In an earlier chapter I wrote about Scone airport and the beginning of the war. In those autumn days of 1939 we once stayed a night on our way to Scone with Charles Coulson, who was then in Dundee. Later he became a distinguished professor of theoretical chemistry in Oxford, and I had immediate sympathy for his form of liberal theology. He held that the methods of science and religion had much in common, that the scientist's experience went far beyond the laboratory to include a sense of reverence and humility and an awareness of beauty and order. How is it possible for any being, whether or not he is a scientist, to look toward the clear night sky and see nebulae, by the light that left them before our earth came into existence, without a profound feeling of humility? I agree with Coulson that we make sense of our relationship with our fellowmen and with the universe "only in terms of a God—partly seen in science, and in art and history and philosophy, partly experienced in wholly personal terms."

Perhaps as I age I shall agree more and more with Paul Tillich that God is Being itself—not the supreme transcendent Being of Barth's theology who is known only to man through self-disclosure. For Tillich, faith was the state of "being grasped by that which concerns one unconditionally." When we are deeply moved by an event, by a Bach mass, or by the radiance of the sunset, we have faith. The events belong to this world, the mass is man created, and the sunset is a terrestrial phenomenon, but their ethos is beyond the world, for no mathematical tools can reveal the beauty of the mass and no spectroscopic analysis can reveal the beauty of the sunset.

My sympathies, therefore, are not with the theologies that deny that religion and science interact. I reject the fundamental thesis of neo-orthodoxy, existentialism, and linguistic analysis, because I believe that civilization is in urgent need of a progressive theology that takes cognizance of the ethical problems of the modern world and of the discoveries of the scientists. For me the critical importance of Whitehead's process philosophy is the development of a metaphysical system relevant to both science and religion. And Whitehead's theology echoes much of my upbringing, particularly in his portrayal of a God of persuasion and love and not a God of coercion and power. His action is by evocation of a response.

The Personal Factor

When I was a very young boy, my father advised me not to be tortured by the thought of hell as he had been throughout his youth. He had been born into a community that accepted as axiomatic that God and the angels were in heaven awaiting the saved and that the devil and hellfire awaited the souls who had failed to achieve salvation. For him, heaven was above and hell was below. It was the symbolism powerfully established by Dante in *The Divine Comedy*—the translation of the geocentric cosmology in terms of heaven and hell.

In this epic work Dante enshrined the Aristotelian and epicyclic universe, the texts of the Bible, and the theological reasoning of Aquinas into a seemingly impregnable structure of beliefs, hopes, and fears that dominated the Western world for three centuries. Indeed, the symbolism of the sacred universe pervaded Western thought long after the ideas of Aristotle's physics and the epicentric universe had been abandoned. It was my father's hope that I would not be troubled throughout my life by such a powerful symbolism. Not troubled, perhaps, but nevertheless as intuitively aware of hell as underneath and heaven above as of the diurnal movement of the sun across the sky. For me *The Dream of Gerontius* is not only a masterpiece by Sir Edward Elgar. The devils' chorus of this oratorio symbolizes the abyss into which I could have been thrown but for the fortunate influences that have shaped my life and thought.

Nearly all the scientists of my acquaintance have acquired a substantial amateur interest and authority in activities that have little to do with their professional discipline. In this respect they are normal human beings differing little from the members of any other profession. Perhaps I have been especially fortunate in that throughout all my professional life—apart from the years of war—I have engaged in three activities that are far removed from the workbench.

In my youth in the West Country there were many families of Lovells in the surrounding villages, and they were nearly all accomplished musicians. Fortunately I inherited their interest. As a boy I was taught to play the piano, but it was the organ that soon became the practical outlet of this interest. During my years in the University of Bristol I studied the organ with Raymond Jones, a friend in the city of Bath who was then the organist at a church in that city. Later he became the organist at Bath Abbey and still plays occasionally there as organist emeritus. In my fifties I studied again with the brilliant young organist of the University of Manchester—Brian Runnett, who in 1970 was tragically killed in a car accident while driving home after a recital he had given in Westminster Abbey. These musical interests have led to associations with people and organizations that have no connection whatsoever with Jodrell or with science. The music societies in four of the nearby cities or small towns claim my attention, and for twelve years I was president of the Guild of Church Musicians. In London in the 1960s I was elected to the Court of the ancient Worshipful Company of Musicians and by 1986–87 had progressed to be the master of this city company, which has existed for nearly five hundred years. Like those of the other ancient city companies, the traditions and activities of the Company of Musicians are far removed from the normal life of the scientist. The banquets, dinners, and regalia are redolent of a past age that one enters as if in a time warp from the twentieth century.

Then there is the particularly English avocation of the game of cricket. This game has never grasped the globe, but my occasional references to the game in this book indicate the hold it has had on me since boyhood. Even in the most despairing moments during the building of the telescope, the anxieties and worries would vanish for a few hours each week on the cricket field. There have been occasional links between the game and my work, but they have been few and memorable. One of these occurred when our Soviet colleagues were using the telescope to search for their lost space probe to Venus. I told them during the course of a Saturday morning that they were welcome to continue the search in the afternoon but that I must go to the cricket match. Alla Massevitch, the leader of the group, announced that she had been instructed to stay with me during the visit and therefore would accompany me to the cricket field. She did so, but it rained and we waited and waited in the pavilion in the hope that it would cease and we could commence the game before darkness fell. We were not able to play, and Alla Massevitch returned to Moscow with the story of the strange group of Englishmen in white clothes, led by the director of Jodrell Bank, who spent an entire afternoon gazing at the leaden sky and in animated discussion about the effect of the rain on the field of play.

The third of my diversions has also been that of many Englishmen—gardening. It was one of the critical events in my life soon after the war that transformed this interest into a major activity. At the end of the war we had purchased a pleasant house in a suburb of Manchester. It was only a few miles from the university where I had expected my life to be centered. At that time I had never heard of the region called Jodrell Bank, but before we had lived one year in this house my journeys there became far more frequent than those to the university. Early in 1948 the inevitable happened. We had talked increasingly of moving to the open country and nearer to the place that now absorbed my scientific interest. One morning in May fate decreed that the newspaper lay open to reveal that a house called The Quinta in the village of Swettenham was to be auctioned that afternoon. We drove there—a few miles of leafy winding lanes south of Jodrell—and a few hours later the house and several acres of pasture land were ours. Now, more than forty years later, the land is covered with mature trees and shrubs, many grown from seed collected during our overseas visits. Always when I travel there are envelopes for seed and music to practice in case an organ is at hand.

The village of Swettenham, in which we have created these gardens, was, and remains today, an isolated community of fewer

356

than 250 people. When we arrived in 1948 there was no electricity or gas supply or main drainage. There was a village school and a local post office, both of which later disappeared. The life of the farming community was centered on the local public house, the club, the school, and the church. The church was an ancient and beautiful building with a list of rectors extending back to the early fourteenth century. Our new house had once been a rectory, and now it was to become our family home. The last two of our five children were born there, and there in 1987 we were to celebrate our golden wedding anniversary with them, their spouses, and nearly all the thirteen grandchildren. Once when we were in Moscow, the Russians told my wife that the mother of six children is made a heroine of the Soviet Union and that although we had only five children the telescope counted as the sixth and she could therefore be a heroine of the Soviet Union. Although said lightly, that sentiment was perceptive of those who had been at the real core of my life.

The Quinta has been a magnet, the kind of house that it is hard to leave and that pulls one back whenever one is away. I doubt if anyone could remain unaffected by this environment. Indeed, a kind reviewer once wrote that I had composed the Reith Lectures "over the spade." That may well have been a percipient remark, for after that move in 1948 I became surrounded by a village community of simple faith, like the one into which I had been born and in which I had lived until I left Bristol in 1936. God blessed the plow in January and was thanked for the harvest in September. Once more, after an amorphous gap of twenty years, the discipline of the church began to exert its grip—for me the organ to be played on Sunday and for my wife the altar flowers and the decorations for the church festivals soon set simple routines with ramifications of a complexity that only those who have lived in a small community will appreciate. That is how our life has progressed and how the environment of our home life has expanded so that it has become a place of magic for the young and, for us, the place where the unending changes of flora and fauna provide, as Francis Bacon wrote of gardens nearly four centuries ago, the purest of human pleasures.

When I walk into these gardens, I can see, rising a few miles away across the flat Cheshire countryside, the great telescope that has so dominated this account of my scientific life. The telescope still probes the heavens, but, like Job, I feel that the search relates to that "which maketh Arcturus, Orion and Pleiades, and the chambers of the South. Which doeth great things past finding out. . . ."

Glossary

AI Air-Interception Radar, the radar equipment in night fighters used for the detection of enemy aircraft.

AIF The code used in TRE to describe a radar in a night fighter that could "lock" the antenna onto the radar echo from the target and ultimately lead to "blind firing"—that is, destruction of the target without visual contact.

angular diameter The angle subtended at the eye by a distant object. For example, the moon subtends an angle of 0.5 degrees and so does the sun (hence the occurrence of total eclipses), but an object the size of the moon placed at the distance of the sun would subtend an angle of only 6 seconds of arc. Angular-diameter measurements are used here in connection with galaxies that are remote in the universe and where the objects although of great size are so distant that the angle subtended may be only thousandths of a second of arc. If the distance and the angular diameter are known, the size of the object can be determined by simple trigonometry.

AORG The Army Operational Research Group, established at West Byfleet, in Surrey, during World War II.

aperture synthesis telescope An instrumental development in radio astronomy to achieve high resolving power. For example, the 250-foot aperture radio telescope described in this book has a resolving power (beam width) of about 1 degree when working on a wavelength of 1 meter. The aperture would have to be increased to 140 miles to obtain a resolving power of a second of arc on this wavelength. Although it is impossible to build a radio telescope with a reflecting bowl of this size, one can obtain the equivalent resolution by using a number of smaller telescopes to "fill in" the aperture of the equivalent large telescope by making large numbers of observations over many days. The data are stored in a computer, and appropriate analysis makes it possible to synthesize the radio image that would have been obtained by means of a single large aperture radio telescope.

ASV Air to Surface Vessel, the radar equipment in aircraft used for the detection of ships and surfaced submarines.

azimuth The position around the horizon. In the Northern Hemisphere the convention is that due north is 360 degrees (or zero), east 90 degrees, south 180 degrees, and west 270 degrees. The radio telescopes described here can move independently in azimuth and elevation (horizontal = zero degrees, zenith = 90 degrees); this is known as an altazimuth mount.

broadside array A flat array, usually of dipoles (q.v.), used in radio astronomy to obtain a narrow beam of reception. The length of the dipoles and their spacing depends on the wavelength. Although cheap and easy to construct, it has the disadvantage, compared with the paraboloid (dish-type) antenna, that the array will work only on the specific wavelength for which it is designed. In the paraboloid only the single feed at the focus has to be changed for differing wavelengths of operation.

centimetric AI A radar air interception system for night fighters that was developed in the early years of World War II and in which the wavelength of the radar was generally 10 centimeters or less. This enabled a narrow beam of radar waves to be transmitted from the aircraft and thereby avoid the confusing ground returns (q.v.).

CH Chain Home, the network of radar stations built along the east and south coasts of England for the detection of enemy bombers. It played a vital role in the Battle of Britain.

cloud chamber An apparatus devised by the Cambridge physicist C. T. R. Wilson in 1911 in which the track of electrons or other ionized particles can be made visible. If moist air becomes supersaturated, water droplets condense on dust particles to form a mist. Wilson discovered that condensation also occurs on ions, such as alpha-particles, or electrons. In a typical form of the apparatus moist air is enclosed in a chamber and the cooling to produce supersaturation is achieved by suddenly increasing the volume (such as by the release of a piston). With appropriate side illumination, coincident with the expansion, the tracks can be photographed through a solid glass front to the chamber. The apparatus was widely used for investigating the charged particles in cosmic rays and atomic physics but was largely superseded by other methods after World War II.

cosmic-ray ionization Although the term "cosmic ray" is still current, it is now known that the cosmic rays incident on the earth's atmosphere from space are almost entirely charged particles of very high energies. Ninety percent or more of the highest-energy cosmic rays incident on the earth's atmosphere are protons, probably originating in supernova explosions or pulsars. Flares on the sun are also a source of protons of lower energy. When these high-energy particles collide with the atoms of the earth's atmosphere, several types of ionized particles are created. At ground level large numbers of these secondary particles are observed—mainly electrons and some heavier ionized particles, including a few percent of the highly penetrating muons (whose mass is 207 times that of the electron). About five muons pass through our heads every second.

dipole A rod-like antenna used for transmitting and receiving radio waves. A common form is known as the half-wave dipole, in which a rod one-half the length of the wavelength is split at the center point and connected by cable to the transmitter or receiver. This type of antenna has a broad beam of reception. (See also *Yagi aerial.*)

Doppler effect The change in the observed frequency (or wavelength) of sound waves, or of electromagnetic waves (e.g., light or radio waves), when the source of the waves and the observer are moving either toward or away from each other. In the former case the frequency increases (increase of pitch, for sound); in the latter case it is decreased. (See also *redshift.*)

DSIR Department of Scientific and Industrial Research in the United Kingdom. One division of the department was responsible for administering government funds for fundamental research. These aspects of the work of the DSIR were taken over by the newly formed SRC (q.v.) in 1965.

eV (electron volt) A small unit of energy commonly used in atomic physics. It is defined as the energy acquired by an electron in falling through a potential difference of 1 volt ($1 \text{ eV} = 1.60 \times 10^{-19}$ joules).

Faraday effect Electromagnetic waves consist of a periodic electrical disturbance and a periodic magnetic disturbance at right angles to one another—the vibrations being in a plane perpendicular to the direction of propagation. These are described as the electric and magnetic vectors of the electromagnetic wave. When emitted from a single oscillatory source, the wave is "plane polarized"; that is, the electric vector is confined to a single plane. The *Faraday effect* is the rotation of this plane of polarization when the wave passes through an ionized medium. The rate of this rotation depends on the density of the ionized medium, the magnetic field, and the frequency of the waves; the correct formulation was first given by Faraday. In chapter 11 the radio waves transmitted and reflected from the moon were plane polarized, but the plane of polarization was rotated by the passage of the radio waves through the ionosphere in the presence of the earth's magnetic field.

flare stars A class of stars discovered in 1948 by W. J. Luyten. They are red dwarf stars whose temperature, size, and mass are lower than the sun's and that exhibit occasional unpredictable and abrupt increases in brightness. The investigations referred to in chapter 20 revealed that the optical flares are associated with outbursts of radio emission and that the energies involved in these flare phenomena are considerably greater than those in the well-known solar flares.

Fresnel zones The rhythmic variations in brightness in the shadow cast by the edge of an obstacle placed in a beam of light. Known as diffraction, the phenomenon was interpreted in terms of the wave theory of light by the French physicist A. J. Fresnel (1788–1827). Chapter 8 describes an analogous phenomenon, occurring when radio waves are reflected from the ionized meteor trail as the meteor passes the perpendicular from the radar to the trail. In this case the rhythmic variations in the strength of the radar echo were used to determine the velocity of the meteor.

gain A measure of the directional properties of an antenna—a quantity comparing the power radiated (or received) with the power radiated (or received) by a point source radiating (or receiving) isotropically. A single dipole (q.v.) does not radiate, or receive, along the direction of the rod and has a gain of 3/2 over an isotropic radiator. As the size of the antenna increases, the beam of radiation or reception narrows and the gain increases. For a single parabolic-dish type of antenna the gain is proportional to the area of the aperture and inversely proportional to the square of the wavelength.

Gee navigational aid A navigational and bombing aid first used operationally by the Royal Air Force over Germany in March 1942. Three widely separated transmitters in England transmitted pulsed radar signals simultaneously. The center station acted as the "master" and the other two as "slaves." A receiver in the aircraft measured the difference in time of receipt of the pulses from each slave and the master and thereby determined its distance from the master and each slave. The locus of all points at which a constant time difference is observed between the pulse from the master and one of the slaves is a hyperbola. Special charts in the aircraft of these sets of intersecting hyperbolae enabled the navigator to determine the position of the aircraft. The range is limited by the earth's curvature because the aircraft receives the direct ray from the transmitter. With the transmitters on the east coast of England, the maximum range of about 350 miles encompassed the Ruhr. At this range the position of the aircraft could be calculated to within an elliptical area of about 6 miles by 1 mile.

ground returns The signals scattered from the ground by a radar in an aircraft. In the early years of the war the radar in the night fighters used antennae transmitting in a broad beam. The signals scattered from the ground obscured the signal scattered from the target and thus limited the useful range to the height at which the fighter was flying. The development of narrow-beam centimeter equipment overcame this difficulty, and in 1942 it was found that a narrow radar beam on a wavelength of 10 centimeters or less gave enough discrimination in the ground returns to produce a useful map. This finding led to the development of the H_2S bombing aid described in chapter 6.

H_2S The centimeter navigational and blind-bombing radar system developed for use by the night bombers of the Royal Air Force in World War II.

H_2X The American version of the British 3-centimeter H_2S system.

Hubble constant In 1929 Edwin Hubble discovered that the redshifts (q.v.) of the spectral lines in the galaxies were proportional to the distance of the galaxy. It is generally accepted that the redshift is a Doppler effect associated with the recession of the galaxies in an expanding universe. Out to great distances the relation between the redshift (velocity of recession) and the distance is linear. Relativistic effects modify the linear relation for velocities of recession approaching the velocity of light. The Hubble constant expresses the relation between the velocity of recession and the distance. The inverse of the Hubble constant is known as the Hubble time and, on the assumption of the linear relation, gives the age of the universe since the beginning of the expansion. At present there is dispute about the value of the Hubble constant, but there seems general agreement that it lies between 50 and 100 kilometers per second per megaparsec. The former value gives the age of the universe as 20 billion years and the latter as 10 billion years.

IAU The International Astronomical Union. Its general assemblies, convened in different countries of the world every three years, form a major meeting point for the international astronomical community.

ICBM Intercontinental ballistic missile.

IGY International Geophysical Year. The IGY referred to in this book was convened by the International Scientific Unions for a worldwide research in geophysics and related subjects during the 1957–58 period of maximum solar and sunspot activity.

interferometer A special form of radio telescope. The resolving power of a single radio telescope at a given wavelength depends on its aperture. Resolving powers equivalent to very large apertures can be obtained by the use of two small-aperture radio telescopes connected to a common receiver either by cable or by radio link. The simple two-element interferometer gives the angular size across one line of the radio source. Aperture synthesis (q.v.) provides a more sophisticated means of combining pairs of interferometric patterns so that a more complete picture can be deduced of the structure of the source.

klystron A form of vacuum tube developed in the United States in the 1930s for the generation and amplification of microwaves. The klystron was widely used when radars working on wavelengths of 10 centimeters or less were developed in World War II. A beam of electrons is "bunched" by means of an alternating voltage applied to grids in the tube. A system of resonant cavities enables microwave power to be obtained with a frequency equal to that of the bunching.

Leigh light In the early years of World War II the German U-boats tended to remain submerged during the daytime but to surface at night to recharge their batteries. The development of ASV radars enabled the night flying aircraft of the RAF Coastal Command to detect the U-boats in darkness, but, except under occasional conditions of favorable moonlight and good visibility, they could not carry out attacks, because the U-boat could not be positively identified. In 1942 a powerful searchlight with a flat-topped beam was installed underneath the fuselage of the radar-equipped Coastal Command night patrols. The searchlight was mounted in a retractable cupola, and during the final approach to a target detected by radar it was lowered and the target illuminated so that a visual identification could be made. The device was known as the Leigh light, after its inventor, Wing Commander H. Leigh of RAF Coastal Command.

libration The period of the moon's rotation on its axis (27.3 days) is the same as its period of revolution around the earth, and therefore we see the same face of the moon at all times. However, because the moon's orbit is slightly elliptical, we see the moon sometimes ahead of and sometimes behind its mean position. This apparent "wobble," known as libration (in longitude), enables us to see 59 percent of the moon's surface, instead of only one-half if the orbit were circular. There is also a libration in latitude because the moon's rotation axis is displaced from the plane of its orbit by 6 degrees 41 minutes. This allows areas of the moon's surface beyond its north and south poles to be seen from the earth.

magnetron The special form of magnetron that became very important in many radar devices used in World War II is known as the *cavity magnetron*. The underlying principle was discovered by J. T. Randall and H. A. Boot in the University of Birmingham and was quickly applied to the development of a powerful generator of microwaves in the centimeter wavelength region. The cavity magnetron consists of a copper block in which cylindrical cavities are drilled. Electrons produced from a cathode in the center of the block are accelerated toward the cavities, but a magnetic field aligned with the axis of the block forces the electrons into spiral paths. The electromagnetic fields from the successive cavities bunch the stream of electrons, and oscillations are stimulated in the cavities, which can be extracted to form a powerful microwave signal.

metadyne control system A form of rotary-power amplifier using direct-current machines. Thus, from small power inputs provided from an electronic circuit, electric motor driving systems involving many kilowatts of power can be produced. The reference here is to the driving of the telescope in azimuth and elevation where the direct-current driving motors, totaling 200 horsepower in each motion, are controlled by the small voltages produced by the electronic devices that indicate the extent to which the actual position of the telescope differs from the required (demanded) position. The form of servo control eventually used in the telescope described in this book is known as a Ward-Leonard system, differing from the metadyne system in the details of the rotary-power amplification.

microwave background radiation Discovered in 1965 by A. A. Penzias and R. W. Wilson of the Bell Laboratories, in New Jersey, the radiation is believed to be the relic radiation from the hot, dense initial state of the universe that was emitted shortly after the beginning of the expansion (the "big bang"). While testing a sensitive microwave antenna and receiver system for communication via earth satellites, Penzias and Wilson found a background signal of very low intensity that was uniform from any direction in the sky. The strength of this radiation is equivalent to that emitted by a blackbody at a temperature of about 3 degrees Kelvin (-270 degrees centigrade) and is therefore often referred to as the 3-degree blackbody radiation or cosmic background radiation. The radiation is believed to have been emitted between 10 and 20 billion years ago, when the universe had been expanding for less than a million years and had cooled to a few thousand degrees Kelvin. Subsequently the radiation became redshifted with the expansion of the universe and is now observed with the peak of intensity in the millimeter wavelength region with a temperature of 2.7 degrees Kelvin. The discovery and subsequent investigation of this microwave background has been of cardinal importance in cosmology.

neutron star A star composed almost entirely of neutrons formed as a normal star collapses when its hydrogen is exhausted. Stars with masses equal to or less than the sun's will collapse into a white dwarf state, but for stars of mass more than about 1.4 times the sun's the gravitational forces are so great that the star collapses to the neutron state, in which the entire mass of the star is compressed with a diameter of only a few kilometers. The density of a neutron star is immense—some billion tons per cubic centimeter. The pulsars discussed in chapter 20 are identified as neutron stars that are the end product of the collapse of a star in a supernova explosion.

Oboe An accurate bombing aid first used operationally by the Royal Air Force in December 1942. Two radar stations on accurately surveyed positions on the east coast of England controlled an aircraft that carried a beacon transmitter so that it could be guided to fly at a constant distance from one of the ground stations (the "cat" station). Thus the aircraft flew in a circle with the cat station at its center. The radius of the circle was chosen so that the aircraft flew over the target.

364

The other ground station (the "mouse") also measured the range from the beacon carried in the aircraft and transmitted the "release" signal to the aircraft at the calculated time. As in the case of Gee, the range was limited because the aircraft had to observe the direct radar transmission from the ground stations. Although the accuracy was very high—at a range of 250 miles an aircraft flying at 30,000 feet could release its bombs to within 120 yards of a selected spot—only one aircraft could be controlled every ten minutes. However, the use of high-flying Mosquito aircraft fitted with Oboe to mark the target by dropping flares led to great devastation in the Ruhr.

occultation The obscuration of a distant body by a nearer one. The occurrence of a total eclipse of the sun is an occultation of the sun by the moon. In this book reference is made to the observation of the change in intensity of the radio waves from a remote galaxy as it was occulted by the moon—an observation that led to an accurate determination of its position.

PAC The Public Accounts Committee, comprising members of Parliament charged with the task of investigating cases of overexpenditure of public money.

particles (fundamental) Scientists refer to fundamental particles as those forming the ultimate constituents of matter. Formerly, atoms were regarded as fundamental particles, but in this century atomic theory has undergone radical and continual revision. In the years covered by the early chapters of this book, atoms were discovered to be composed of combinations of electrons and protons, which for some time were regarded as the fundamental particles. Then James Chadwick discovered the neutron (a particle with the mass of the proton but no charge), then there arose the concept of the neutrino (mass of the electron but no charge), and since World War II the investigations of the physicists with the powerful particle accelerators have led to the discovery of a multitude of particles of various masses, charges, stability, and other properties. Recent theoretical developments have led to the idea of the "quark" as the really fundamental particle of the universe—which, in combination or in interactions, leads to all the particles formerly regarded as fundamental.

particles (primary and secondary) The description of particles as primary or secondary in this book refers to the cosmic-ray particles (see cosmic-ray ionization). The primary particles are those of high energy incident on the earth's atmosphere from sources in the universe—believed to be composed largely of protons. When these interact with the atoms of the earth's atmosphere, large numbers of secondary particles are produced. At sea level these are mainly electrons; in addition, there is a small proportion of heavier, very penetrating, ionized particles.

perspex cupola The perspex housing mounted underneath the fuselage of bomber aircraft to house a radar scanner. The word "cupola" is normally used to describe a dome. The objects under the fuselage of the bomber aircraft were strictly inverted cupolas. Also, they were elongated and not spherical (see figure 2).

polar diagram The shape of the beam of radiation emitted (or received—the two are identical). The beam width is usually defined as the width of the beam when the intensity is one-half the maximum. There are alternative methods of picturing the shape of the beam. In the conventional polar diagram the antenna is the "pole," or origin, of the beam. The radiated pattern normally consists of a prominent forward lobe with a number of (unwanted) minor or side lobes.

radiant point of meteor shower When the earth crosses the orbit of a comet, large numbers of small particles associated with the comet burn up in the high atmosphere of the earth, giving rise to meteors, or "shooting stars." Although these particles enter the atmosphere in parallel paths, when viewed by an observer on earth they appear to diverge from a point in the sky because of the effects of perspective. This point is known as the radiant point of the meteor shower and is named after the constellation of stars in which it appears to lie. Hence, for example, the meteors of the Perseid shower appear to radiate from a point in the constellation of Perseus. The nomenclature is, of course, merely a convenient reference, since the meteors are confined to the Solar System and have no connection with the distant stars forming the constellation of Perseus.

radio stars The term was originally used in the late 1940s and early 1950s to describe the "discrete" or "localized" sources of radio emission that did not appear to have any visible counterpart. For some time they were believed to be a hitherto unknown type of stars—"dark stars" emitting radio waves in the Milky Way. In the 1950s these radio-emitting objects were discovered, on the contrary, to be distant objects in the universe and became identified as radio galaxies, quasars, etc. Later, certain types of stars in the Milky Way were discovered to be radio emitters—the sun emits radio waves and so do the flare stars (q.v.). However, these and other types of stars from which radio emission has been detected are in the catalogs of stars visible in the optical telescopes, and the term "radio star" is no longer used to describe a particular type of object not otherwise identifiable.

RAS The Royal Astronomical Society of London.

red dwarfs The coolest visible stars known. Whereas the surface temperature of the sun, regarded as an average type of star, is 6000 degrees Kelvin, that of the red dwarfs is half or less of this value. Red dwarfs are therefore red and of low luminosity, being visible only as telescopic objects. They are also much less massive than the sun—in some cases having a tenth or less of the solar mass. Because of their faintness only a few hundred of the nearest have been cataloged, but they are believed to account for perhaps 80 percent of the stars in the galaxy. See also *flare stars*.

redshift The lengthening of the wavelength (that is, toward the red end of the spectrum) observed in the spectra of the distant galaxies. The redshift of the spectral lines from certain astronomical objects was first observed in 1912 by V. M. Slipher at the Lowell Observatory, in Arizona. The discovery in the 1920s by Edwin Hubble that the objects exhibiting the redshift were external to the Milky Way led to the interpretation in terms of the Doppler shift (q.v.) associated with the expansion of the universe (see also *Hubble constant*).

scintillation (interplanetary) The scintillation (twinkling) of the stars is a well-known phenomenon, caused by the uneven refraction of the light from the star as it passes through the earth's atmosphere. An analogous scintillation, or variation in strength of the signals, is observed in the radio wave part of the spectrum. Two types are distinguished. One arises when the radio waves from a distant object traverse the ionized regions of the earth's atmosphere. The other arises when the radio waves from a distant source traverse ionized clouds (plasma) in interstellar space. This interplanetary-scintillation effect is particularly prominent when the signals from the radio source pass through the extended corona of the sun. The period of the fluctuations is generally on the order of seconds, and it was during the course of these investigations that pulsars were discovered (see chapter 20).

searchlight aerial One of the first directional aerials (radio telescopes) built at Jodrell Bank. This consisted of an array of Yagi aerials (q.v.) built onto a borrowed army searchlight mount. This searchlight aerial is described in chapters 7 and 8.

servo system A device used to convert a low-powered mechanical motion into one requiring considerably greater power. The output power is generally proportional to the input power, which is often generated by electronic means. The metadyne (q.v.) and the Ward-Leonard systems used in the telescope are servo systems in which the servo loop is closed, so that the driving system constantly seeks to reduce the error between the required (or demanded) position and the actual position. Closed servo-loop systems are common features of automatic-pilot control of aircraft or ships.

solar parallax The angle subtended by the radius of the earth at the sun. This angle determines the distance of the sun from the earth and is of fundamental importance in astronomy since the earth–sun distance forms the baseline for the measurement of the distances of the nearer stars and, hence, the distance scale for the universe. The significance of the radar detection of the planet Venus in the determination of the solar parallax is discussed in chapter 19.

SRC (SERC) The Science Research Council was formed in 1965 to take over from the DSIR (q.v.) the allocation of government funds for fundamental research and to assume responsibility for other institutions financed by government departments (e.g., the Royal Observatories). In April 1981 its name was changed to the Science and Engineering Research Council.

supernova A stellar explosion that occurs at the end of the life of a massive star. The mass limit at which a star undergoes such a violent disruption is about 1.5 times greater than the mass of the sun. When a star ends its life in a supernova explosion, the luminosity may increase by some ten billion times. The object in the Milky Way known as the Crab nebula is a famous case of a supernova, observed by Chinese astronomers in the year A.D. 1054. The gaseous remnants are readily visible in large telescopes, and the core into which the major part of the original stellar mass collapsed is a neutron star (q.v.), observable as a pulsar (see chapter 20).

transient echoes The radar echoes from the ionized trails of meteors (shooting stars). The duration of the majority of the echoes observed with the meter wavelength radars described in this book was a fraction of a second.

transit telescope In general a fixed telescope used to observe the precise moment when a star crosses (transits) the meridian. In this book the term applies to the first large radio telescope fixed to the ground so that the narrow beam was directed upward (to the zenith). Only the radio waves from the region of the sky passing through the zenith as the earth rotates could be observed. This Jodrell transit telescope is described in Chapter 9.

TRE The British wartime radar establishment. The original prewar radar research department was at Bawdsey Manor, on the east coast of England, and was known as the Bawdsey Research Station (BRS). When war broke out, the establishment moved to Dundee, in Scotland, and became the Air Ministry Research Establishment (AMRE). In the spring of 1940 this establishment moved to the Dorset coast, in the south of England. With a change in ministerial arrangements, AMRE became MAPRE (the Ministry of Aircraft Research Establishment); finally, in November 1940, it became TRE (Telecommunications Research Establishment). In the spring of 1942 TRE was evacuated to Great Malvern, in Worcestershire, and in the postwar years various mergers and further changes of name took place. It is now (1989) the Royal Signals and Radar Establishment (RSRE).

URSI Union Radio-Scientifique Internationale, or International Scientific Radio Union. Its general assemblies provide the meeting ground for the several users of radio techniques in scientific research.

V-1 and V-2 *Vergeltungswaffen*, or vengeance weapons. The V-1 was the German pilotless 5,000-pound airplane carrying a ton of high explosives. The bombardment of London and the southeast of England with these weapons began in June 1944 and ended with the capture of the launching sites by the Allies in September 1944. The V-2 was a ten-ton ballistic rocket carrying a ton of high explosives over a range of 180 to 210 miles. The attacks on London with the V-2 began on 9 September 1944, a few days after the end of the V-1 attacks, and ended with the capture of the launching sites by the Allied armies on 27 March 1945.

wave band A range of wavelengths in the radio wave region of the electromagnetic spectrum. At ground level the range of wavelengths over which it is possible to receive radio waves from space extends from about 1 centimeter to 50 meters. The lower end is limited by the absorption of incoming radiation by the earth's atmosphere. On dry and high mountain sites the limit can be extended to a wavelength of a few millimeters. The upper end is limited by the scattering and absorption of the incoming radiation in the ionosphere surrounding the earth at altitudes of about 100–400 kilometers. This waveband is widely used for communications (radio and television) and for many other commercial and defense interests. The precise wavelengths for such use are allocated by regional and international bodies. A number of wavelengths in this range are allocated to radio astronomy. For many uses a narrow band of wavelengths—usually specified as the bandwidth—is defined. The wavelength (λ) is inversely proportional to the frequency (f) which is generally used, particularly in scientific work. If c is the velocity of light, then $c = \lambda f$, and thus, with λ measured in meters and f in megahertz (hertz is cycles per second), $\lambda = 300/f$.

wind loading A term used here to refer to the pressure of the wind on the telescope structure. When the large telescope was designed, the effect of wind pressure, and particularly of gusty winds, on open steel structures was not easy to calculate; assessments were often made by means of wind tunnel tests on a scale model. The variation of wind with height above ground and the effects of wind on the curved paraboloidal bowl of the telescope were difficult to determine. Under gusty conditions the relation of the frequency of the gusts to the natural frequency of the structure is of critical importance. The destruction of the Tacoma Bridge in Washington State during gusty winds illustrates the critical value of this relation.

Yagi aerial A device for obtaining a narrow beam for the reception or transmission of radio waves, developed by the Japanese engineer Hidetsugu Yagi in 1928. This consists of a half-wave dipole usually backed by a single rod acting as a parasitic

368

reflector and a number of rods in front of the dipole acting as parasitic directors. This type of directional antenna has been widely used, mounted on housetops, for television receivers. The searchlight aerial (q.v.) built at Jodrell Bank used an array of Yagi aerials (see chapters 7 and 8).

Zenith The point immediately over the observer's head. Conventionally the horizon (horizontal) is taken as zero degrees, and the zenith is then at 90 degrees. Thus, in this book, where reference is made to the telescope's pointing toward the zenith, the beam is vertically directed. In exact astronomical work the zenith has to be defined more precisely, since the earth is not a perfectly uniform sphere.

Index

371

INDEX

INDEX

INDEX

379

INDEX